Information Technology Ethics

This book explains moral dilemmas in the IT industry, focusing on a different area of IT ethics. It starts by introducing theories, decision-making models, and the fundamental function of moral leadership in IT companies, followed by ethical obligations related to intellectual property, privacy, professional ethics, and software development. It looks at how IT affects society in terms of accessibility and well-being, including cybersecurity tactics, privacy and data protection legislation, legal and regulatory frameworks, and the moral obligations of IT workers and companies.

Features:

- Includes extensive examination of IT ethics encompassing fundamental theories.
- Covers foundational concepts, intricate ethical quandaries, and societal ramifications.
- Provides practical guidance with real-world illustrations, case studies, and optimal approaches.
- Integrates cultural perspectives, case studies, and discussions on worldwide legal frameworks, accessibility challenges, and environmental sustainability.
- Reviews the Canadian perspective on IT ethics, exploring the specific legal and ethical landscape of the country.

This book is aimed at senior undergraduate and graduate students in computer engineering and IT, as well as professionals.

Computational Intelligence Techniques
Series Editor: Vishal Jain

The objective of this series is to provide researchers a platform to present state of the art innovations, research, and design and implement methodological and algorithmic solutions to data processing problems, designing and analyzing evolving trends in health informatics and computer-aided diagnosis. This series provides support and aid to researchers involved in designing decision support systems that will permit societal acceptance of ambient intelligence. The overall goal of this series is to present the latest snapshot of ongoing research as well as to shed further light on future directions in this space. The series presents novel technical studies as well as position and vision papers comprising hypothetical/speculative scenarios. The book series seeks to compile all aspects of computational intelligence techniques from fundamental principles to current advanced concepts. For this series, we invite researchers, academicians and professionals to contribute, expressing their ideas and research in the application of intelligent techniques to the field of engineering in handbook, reference, or monograph volumes.

Blockchain Technology for IoE
Security and Privacy Perspectives
Edited by Arun Solanki, Anuj Kumar Singh, and Sudeep Tanwar

Convergence of IoT, Blockchain and Computational Intelligence in Smart Cities
Edited by Rajendra Kumar, Vishal Jain, Leong Wai Yie, and Sunantha Prime Teyarachakul

Intelligent Techniques for Cyber-Physical Systems
Edited by Mohammad Sajid, Anil Kumar, Jagendra Singh, Osamah Ibrahim Khalaf, and Mukesh Prasad

Big Data Computing
Advances in Technologies, Methodologies, and Applications
Edited by Tanvir Habib Sardar and Bishwajeet Kumar Pandey

Software Defined Network Frameworks
Security Issues and Use Cases
Edited by Mandeep Kaur, Vishal Jain, Parma Nand and Nitin Rakesh

Machine Learning Hybridization and Optimization for Intelligent Applications
Edited by Tanvir Habib Sardar and Bishwajeet Kumar Pandey

Information Technology Ethics
Hamed Taherdoost

For more information about this series, please visit: www.routledge.com/Computational-Intelligence-Techniques/book-series/CIT

Information Technology Ethics

Hamed Taherdoost

CRC Press
Taylor & Francis Group
Boca Raton London New York

CRC Press is an imprint of the
Taylor & Francis Group, an **informa** business

Designed cover image: www.shutterstock.com

First edition published 2024
by CRC Press
2385 NW Executive Center Drive, Suite 320, Boca Raton FL 33431

and by CRC Press
4 Park Square, Milton Park, Abingdon, Oxon, OX14 4RN

CRC Press is an imprint of Taylor & Francis Group, LLC

© 2025 Hamed Taherdoost

ISBN: 9781032953489 (hbk)
ISBN: 9781032812410 (pbk)
ISBN: 9781003584452 (ebk)

DOI: 10.1201/9781003584452

Typeset in Times
by codeMantra

To my daughter, Hamta, and my son, Kiasha, who remind me of the future we are shaping,

This book is for you, my anchors and my motivation. May your dreams flourish in a world where technology serves humanity with integrity and care. You are my greatest achievement and the heart of all my endeavors.

Contents

Preface...xi
Acknowledgments.. xiii
About the Author xv

Chapter 1   An Overview of Ethics in Information Technology............................1

    1.1   Introduction to Ethical Principles...1
    1.2   Ethical Theories and Their Application in IT.............................2
    1.3   Ethical Decision-Making Models for IT Professionals..............5
    1.4   The Importance of Ethical Leadership in IT Organizations.......8
    1.5   Addressing Ethical Challenges in the IT Industry 10
    1.6   The Role of Ethical Codes and Standards in IT....................... 14
    References 17

Chapter 2   Ethical Responsibilities for IT Workers and Users 18

    2.1   Professional Ethics for IT Practitioners.................................. 18
    2.2   Ethical Obligations in IT Project Management........................20
    2.3   Ethical Implications of IT Outsourcing and Offshoring...........23
    2.4   User Responsibilities and Ethical Use of Technology...............27
    2.5   Ethical Challenges in IT Work Environments 30
    2.6   Ethical Issues in IT Training and Education.............................33
    References 36

Chapter 3   Computer and Internet Crime: Ethical Implications
and Mitigation 37

    3.1   Cybercrime and Its Impact on Society 37
    3.2   Ethical Issues in Hacking and Unauthorized Access............... 41
    3.3   Cybersecurity and Ethical Responsibilities.............................44
    3.4   Ethical Considerations in Data Breaches and Incident
Response.. 48
    3.5   Cyberbullying and Online Harassment....................................51
    3.6   Ethical Approaches to Digital Forensics and Cyber
Investigations..55
    References 58

Chapter 4   Privacy: Ethical Dimensions in the Digital Age 59

    4.1   The Importance of Privacy in the Digital Era...........................59
    4.2   Privacy Rights, Data Protection, and Consent 62
    4.3   Ethical Considerations in Data Collection and Usage 65
    4.4   Balancing Privacy with Security and Public Interest................ 67

　　　　4.5　Privacy Breaches and Data Leaks: Ethical Implications 72
　　　　4.6　Ethical Approaches to Privacy Management in
　　　　　　Organizations 74
　　　　References 75

Chapter 5　Freedom of Expression: Ethical Boundaries in the
　　　　　　Digital Realm 76

　　　　5.1　Understanding Freedom of Expression in the Digital Age 76
　　　　5.2　Ethical Challenges of Balancing Free Speech and Hate
　　　　　　Speech 78
　　　　5.3　Online Disinformation and the Spread of Fake News.............. 80
　　　　5.4　Ethical Implications of Content Moderation and
　　　　　　Censorship 82
　　　　5.5　Online Harassment and Cyberbullying...................................... 85
　　　　5.6　Protecting Free Speech while Upholding Ethical Standards 87
　　　　References 89

Chapter 6　Intellectual Property: Ethical Considerations in the Digital
　　　　　　Landscape..90

　　　　6.1　The Concept of Intellectual Property in the Information
　　　　　　Age.. 90
　　　　6.2　Copyright, Fair Use, and Ethical Decision-Making..................92
　　　　6.3　Ethical Implications of Plagiarism and Content Theft..............93
　　　　6.4　Open-Source Software and Ethical Collaboration....................96
　　　　6.5　Balancing IP Rights with Innovation 98
　　　　6.6　Ethical Practices in Licensing, Patents, and Trademarks 100
　　　　References 103

Chapter 7　Ethical Software Development and Testing 104

　　　　7.1　Ethical Considerations in Software Development Lifecycle ... 104
　　　　7.2　Balancing User Needs, Business Interests, and Ethical
　　　　　　Responsibilities...108
　　　　7.3　Quality Assurance and Ethical Testing Practices.................... 110
　　　　7.4　Addressing Bias and Discrimination in Algorithm Design..... 113
　　　　7.5　Ethical Use of AI and Machine Learning in Software
　　　　　　Development 115
　　　　7.6　Ensuring Ethical Standards in Software Maintenance and
　　　　　　Updates 117
　　　　References 117

Chapter 8　The Impact of Information Technology on Society and
　　　　　　Well-Being.. 119

　　　　8.1　Ethical Implications of Technological Advancements on
　　　　　　Society 119

8.2 Digital Divide and Ethical Considerations of
Accessibility 123
8.3 Ethical Challenges in Technological Innovations and
Disruptions 125
8.4 Ethical Use of Technology for Social Good and
Sustainable Development 127
8.5 Impact of IT on Mental Health and Well-Being 129
8.6 Ethical Responsibilities in Promoting Digital Inclusion
and Equality 132
References 133

Chapter 9 Social Networking and Online Communities: Ethical
Engagement 135

9.1 Ethical Considerations in Building Online
Communities...135
9.2 Privacy and Security Concerns in Social Networking
Platforms..139
9.3 Ethical Implications of Online Influencer Culture.................. 141
9.4 Cyberbullying and Toxic Behavior in Social Media 143
9.5 Ethical Use of Data and User Analytics in Social
Networking 145
9.6 Promoting Digital Citizenship and Responsible Online
Engagement 148
References 150

Chapter 10 Ethics of IT Governance and Policy 153

10.1 Ethical Considerations in IT Governance Frameworks......... 153
10.2 Ethical Decision-Making in IT Policy Development 156
10.3 Ethical Implications of Surveillance and Governmental
Control..159
10.4 Ensuring Ethical Use of IT in Public Sector
Organizations... 161
10.5 Addressing Ethical Challenges in IT Regulation and
Compliance...163
10.6 Ethical Leadership in Shaping IT Policies and
Standards 164
References 166

Chapter 11 Risk, Responsibility, and Ethical Accountability in IT 169

11.1 Risk in IT and Its Ethical Implications................................. 169
11.2 Ethical Responsibilities in Risk Assessment and
Management..171
11.3 Balancing Organizational Goals with Ethical Risk
Mitigation...173

11.4   Ethical Challenges in Data Security and Breach
        Response 175
11.5   Ethical Considerations in Disaster Recovery and Business
        Continuity 178
11.6   Promoting a Culture of Ethical Accountability in IT
        Organizations 180
References 182

Chapter 12 Ethics in Information Technology in Canada 185

12.1   Canadian Legal and Regulatory Frameworks for
        IT Ethics 185
12.2   Ethical Challenges and Considerations in Canadian IT
        Industry 189
12.3   Privacy and Data Protection Laws in Canada 192
12.4   Addressing Ethical Implications of Canadian
        Cybersecurity Strategies 194
12.5   Indigenous Rights and Ethical Use of Technology in
        Canada 196
References 198

Chapter 13 AI Ethics in IT: From Design to Practical Implementation 199

13.1   Fundamentals of AI 199
13.2   Implementation Frameworks for AI Systems 200
13.3   Ethical Considerations in AI Model Training 202
13.4   AI Governance and Compliance 205
13.5   Ethical Challenges in AI Applications 205
13.6   Tools and Technologies for Ethical AI Development 208
13.7   AI Security 211
13.8   Future Directions in AI Ethics and IT 212
References 213

Index 217

Preface

In an age where technology permeates every facet of our lives, the ethical considerations surrounding its development, deployment, and usage have never been more critical. This book, *Ethics in Information Technology: A Comprehensive Exploration*, seeks to provide a thorough examination of the ethical challenges and responsibilities that IT professionals, organizations, and users encounter in today's digital landscape.

As we delve into topics ranging from data privacy and cybersecurity to the ethical implications of artificial intelligence and the responsibilities of digital citizenship, our aim is to present a balanced and comprehensive view of the current state of ethics in information technology. Each chapter is meticulously crafted to address specific areas of concern, offering both theoretical insights and practical guidance to navigate the complex ethical dilemmas that arise in the ever-evolving IT industry.

We explore foundational ethical theories and their application in IT, the ethical responsibilities of IT workers and users, the impact of cybercrime, and the delicate balance between privacy and security. Additionally, we examine the ethical boundaries of freedom of expression online, the implications of intellectual property in the digital age, and the importance of ethical software development. This book also addresses the societal impact of technological advancements, the role of IT governance and policy, and the specific ethical considerations relevant to the Canadian IT landscape.

We hope that this book will serve as a valuable resource for IT professionals, students, educators, policymakers, and anyone interested in the ethical dimensions of information technology. As you engage with the content, we encourage you to reflect on the ethical implications of your actions and decisions in the digital realm, and to strive toward creating a more ethical and just technological future.

Acknowledgments

First and foremost, I would like to express my deepest gratitude to my parents, who were my first and most profound source of learning ethics. Their unwavering values, guidance, and love have shaped the foundation of my personal and professional journey.

I am deeply thankful to my wife, Mitra, for her unwavering support, understanding, and patience throughout the process of writing this book. Her encouragement and belief in me have been a constant source of inspiration.

To my children, Hamta and Kiasha, I am grateful for their love, understanding, and for bringing endless joy into my life. Their presence has provided me with the motivation to persevere through the challenges of this project.

I am also grateful to my colleagues for their collaboration, insights, and camaraderie. Their support and encouragement have been invaluable, and I am fortunate to work alongside such dedicated individuals.

Finally, to everyone who has contributed to this endeavor, directly or indirectly, I extend my heartfelt appreciation. Their support has been instrumental in bringing this project to fruition, and I am deeply thankful for their contributions.

About the Author

Dr. Hamed Taherdoost is an award-winning leader in research and development, known for his contributions across both industry and academia. He is the founder of Hamta Business Corporation, Associate Professor and Chair of RSAC at University Canada West, & Director of R&D at Q Minded | Quark Minded Technology Inc. He has over 20 years of experience in both industry and academic sectors. He has worked at international companies from Cyprus, the UK, Malta, Iran, Malaysia, and Canada and has been highly involved in development of several projects in different industries, healthcare, transportation, residential, oil and gas and IT. Additionally, he has served as a trusted technical and technology consultant for multiple companies, providing advisory and mentorship.

In academia, **Dr. Taherdoost** has held teaching positions in Southeast Asia, the Middle East, Europe and North America since 2009. He began his academic career as a lecturer at AU & PNU and later served as an adjunct professor and supervios at Westcliff University, USA, and Gisma University of Applied Sciences, Germany. His research tenure at IAU lasted over eight years, during which he supervised numerous students.

Dr. Taherdoost has a prolific publishing record, with over 300 scientific articles in top-tier journals and conference proceedings. His work is widely recognized, with a strong citation impact reflecting in a high h-index, and his contributions span book chapters, edited volumes, and authored books focusing on technology and research methodology. His textbooks on E-Business and Digital Transformation are teaching at global institutions, including the University of Malta, the University of Macau, and University Canada West. His research accomplishments have earned him a place among the top 10 SSRN Business Authors since 2022, as well as consistent recognition in the Stanford-Elsevier list of the world's top 2% of Scientists from 2021 to the present.

Dr. Taherdoost has organized and chaired many workshops and conferences and has frequently been invited as a keynote speaker. He is an active member of the editorial, reviewer, and advisory boards for several prestigious journals published by Taylor & Francis, Springer, Emerald, Elsevier, MDPI, EAI, IGI Publishing, and Inderscience. He has also participated as an organising, scientific and technical committee member in over 300 conferences held across Europe, America, Australia, Asia, and Africa.

Dr. Taherdoost's leadership and innovation have been recognized through numerous prestigious accolades and awards at international levels. His achievements span excellence in business, teaching innovation, research methodology, and startup innovation. He has also received multiple honors for his research contributions, including best paper, outstanding reviewer, and editorial excellence awards from renowned journals and organizations.

Dr. Taherdoost is the Editor of a book series with Routledge (Taylor & Francis Group) titled Mastering Academic Excellence: Research, Teaching, Learning, and Publishing. He serves on the editorial boards of several high-impact journals, including the Information Technology Project Management (IGI), Transactions on Scalable Information Systems (EAI), Electronic Government Research (IGI, IF: 1.2), Information Resources Management Journal (IGI), Journal of Blockchain (MDPI), and Data Mining, Modelling, and Management (InderScience). He is also an Associate Editor of Frontiers in Research Metrics and Analytics (Scopus) and Academic Editor of PLOS ONE. He's been a guest editor of special issues in Results in Engineering (Elsevier), Electronics (MDPI), Computers (MDPI), and Discover Computing (Springer).

He is a Certified Cyber Security Professional and Certified Graduate Technologist. He is a GUS Fellow - GUS Institute | Global University Systems, senior member of IEEE, and Working Group Member of International Federation for Information Processing - IFIP TC 11 - Human Aspects of Information Security and Assurance and Information Security Management. Currently, he is involved in several multidisciplinary research projects, including studying innovation in information technology, blockchain, cybersecurity, and technology acceptance.

1 An Overview of Ethics in Information Technology

1.1 INTRODUCTION TO ETHICAL PRINCIPLES

Information technology (IT) ethics are a moral compass guiding individuals and organizations in their technical decisions and actions. It helps distinguish between right and wrong, responsible and irresponsible behavior, and guides navigating the complex IT landscape. These moral principles reflect an individual's character and are crucial in determining how technology will impact society.

The continuously evolving IT ecosystem presents a wide range of ethical issues, including privacy violations, the challenges of artificial intelligence (AI), and digital access disparities. As a result of the rapid development of technology compared with appropriate ethical frameworks, unanticipated moral dilemmas frequently arise and must be resolved with care [1].

At the core of addressing these IT-related ethical challenges lie several fundamental principles.

Privacy and confidentiality involve protecting personal information and handling sensitive data responsibly. IT professionals must implement robust security measures to safeguard user data and maintain trust.

Transparency is about offering clear insights into how technology operates and its potential consequences. Transparency fosters understanding and trust among users, enabling them to make informed decisions about their engagement with technology.

Accountability entails acknowledging the repercussions of technological actions and taking ownership of them. IT professionals must be prepared to answer questions about the impacts of their innovations and practices, ensuring that ethical considerations are prioritized.

Fairness and Equity mean guaranteeing the just distribution of advantages and disadvantages. IT ethics demand that technology benefits all users equitably, avoiding biases and discrimination that could exacerbate social inequalities.

Integrity involves upholding honesty and refraining from actions that compromise data or systems. Integrity in IT practices ensures that systems remain reliable and trustworthy, fostering user confidence and safety.

Access and inclusion are about striving for equal technological accessibility to bridge disparities. Efforts must ensure that technology is available to all, regardless of socio-economic status, physical ability, or geographic location, promoting inclusivity.

Beneficence focuses on leveraging technology for well-being while minimizing potential harm. IT professionals should aim to create positive impacts through technology, enhancing quality of life while vigilantly mitigating risks and negative outcomes.

DOI: 10.1201/9781003584452-1

In IT, making ethical decisions requires recognizing ethical dilemmas, collecting relevant data, analyzing the available options, and coordinating decisions with moral principles. To create a harmonious union between technology and moral principles, the following chapters will investigate several ethical dilemmas and their resolutions.

1.2 ETHICAL THEORIES AND THEIR APPLICATION IN IT

Ethical theories provide a framework for comprehending and analyzing intricate moral dilemmas. In the context of IT, these theories provide valuable insights into the ethical considerations that arise during the design, deployment, and application of technological systems. This section examines prominent ethical theories and their applications to the IT field.

1.2.1 UTILITARIANISM: BALANCING BENEFITS AND HARMS

Utilitarianism, a consequentialist ethical theory, seeks to maximize utility or overall satisfaction. It emphasizes outcomes more than actions and maintains that the moral choice is the one that has the greatest positive impact on the greatest number of people. Utilitarianism provides a useful framework for navigating complex moral quandaries in the discipline of IT.

Utilizing utilitarian concepts is advantageous in every IT situation. When developing software, IT professionals evaluate potential negatives, such as security vulnerabilities, alongside positives, such as enhanced user experience and efficiency. A utilitarian approach to data privacy involves evaluating data insights against user privacy concerns. Ethical AI development seeks to minimize prejudices while maximizing societal benefit. Decisions regarding resource allocation in project management are utilitarian and maximize stakeholder satisfaction. Professionals in IT ensure accessibility for users with disabilities, thereby extending the positive effects of technology.

Although utilitarianism guides moral decisions effectively, it is not without criticism. Some argue that it could lead to disregarding minority interests in favor of majority interests and that measuring utility across diverse individuals can be difficult. Despite objections, utilitarianism provides an effective method for IT professionals to make moral decisions that promote technology's pervasive and beneficial effects on society [2].

1.2.2 DEONTOLOGY: DUTIES AND PRINCIPLES

Deontology is a well-known ethical theory that emphasizes following moral obligations, norms, and principles when making decisions. In contrast to utilitarianism, which emphasizes outcomes, deontology contends that the rightness or wrongness of an action is determined by the activity itself and the principles that motivate it. This concept highlights the importance of adhering to recognized ethical standards and codes of conduct. It comes from the Greek word "deon," which means "duty."

Deontology is a valuable framework for resolving IT-related ethical issues. IT professionals frequently encounter issues relating to user privacy, data security, transparency, and other concerns. The application of deontological principles can aid in making morally sound decisions. IT professionals must, for example, prioritize user privacy, obtain consent before data collection, and ensure secure data storage. Transparency is a second crucial factor. IT professionals should be forthright with users about how technology functions and the risks it poses [3].

However, deontology is susceptible to criticism. Critics assert that it can be rigid and cannot manage complex situations involving competing responsibilities. Moreover, determining universal moral principles can be challenging in the context of technology's complexity and rapid evolution. Despite these challenges, deontology provides IT professionals with a distinct ethical compass. IT professionals can contribute to the ethical development and use of technology by supporting the principles of user privacy, transparency, security, and respect for intellectual property, prioritizing the rights and well-being of individuals in the digital age.

1.2.3 VIRTUE ETHICS: CULTIVATING MORAL CHARACTER

A philosophical viewpoint known as "virtue ethics" emphasizes the development of morally upright characteristics as a paradigm for ethical behavior. In contrast to rule-based or consequence-driven systems, virtue ethics focus on helping individuals cultivate virtues such as honesty, compassion, and integrity. This method emphasizes becoming the person one wishes to be rather than simply deciding what is right or incorrect.

In virtue ethics, the concept of eudaimonia, or human flourishing, attained through moral behavior is essential. Phronesis, also known as practical wisdom, is essential because it enables individuals to practice virtues daily. While acknowledging the role of emotions and habits, virtue ethics emphasize that engaging in virtuous behavior can develop character-enhancing habits.

Virtue ethics provide a framework for making ethical judgments in industries such as IT that transcends mere compliance. IT professionals who practice virtue ethics emphasize characteristics such as accountability, honesty, and compassion. They promote a culture of cooperation and respect by considering the well-being of users and society and the long-term effects. While critics point out the difficulty of uniformly defining virtues, virtue ethics remain effective for IT professionals to align their behaviors with ethical standards, promoting technological advancement and societal development.

1.2.4 RIGHTS-BASED ETHICS: RESPECTING INDIVIDUAL RIGHTS

Rights-based ethics are an essential foundation for ethical decision-making in IT. Respect and protection of individual rights as the governing principles of moral conduct are emphasized heavily in this strategy. These rights, which are inalienable and universal, include privacy, freedom of expression, autonomy, and equitable access. IT professionals are responsible for designing, developing, and implementing technological solutions that respect fundamental rights and defend human dignity.

Autonomy, which permits individuals to make conscious, independent decisions regarding their online experiences, is a cornerstone of rights-based ethics. IT professionals must ensure that users can access their data and manage, delete, and share it with their knowledge and permission. Strong security measures and transparent data management practices make protecting privacy a top priority. Similarly, the right to free speech is safeguarded by preventing excessive censorship or surveillance, promoting open dialogue, and respecting the limits of individual rights.

The development of equitable access is another crucial aspect that compels IT professionals to design inclusive digital ecosystems. This involves adhering to accessibility guidelines so that individuals with disabilities can completely participate. As technology advances, the ethical use of AI and automation is consistent with rights-based ethics by ensuring that these breakthroughs safeguard human rights and do not violate autonomy or privacy. There are still difficulties on a global scale in balancing competing rights and considering cultural differences, which necessitates significant ethical considerations.

In the complex world of IT, rights-based ethics serve as a compass, encouraging the ethical and responsible invention, use, and deployment of technology. By preserving individual rights, IT professionals contribute to a digital environment that respects human dignity, empowers users, and places ethics at the vanguard of technological advancement.

1.2.5   SOCIAL CONTRACT THEORY: COLLECTIVE RESPONSIBILITY

As a fundamental ethical framework, the social contract theory investigates the origin of moral and political obligations in societies. It implies that individuals should unify to form a just and well-ordered community where their rights and interests are protected. This social agreement may be implicit or explicit. This theory, advocated by philosophers such as Thomas Hobbes, John Locke, and Jean-Jacques Rousseau, sheds light on the relationship between individuals, governments, and moral obligations.

IT has significant ethical implications for the social contract theory. By developing technologies prioritizing these values, IT professionals are responsible for protecting individual rights such as privacy and free speech. Governments and regulatory agencies function as digital custodians, establishing and enforcing regulations that ensure responsible technology use and user protection. Transparency, accountability, and responsible digital citizenship require users to actively participate and be informed about data processing procedures, which are crucial components of the digital social compact.

When the concepts of the social contract are applied to the digital world, people, IT professionals, and society work together; this initiative aims to establish an ethical technology environment that places a premium on moral behavior, user rights, and the greater good. The social contract theory, which emphasizes mutual comprehension and the protection of individual interests, guides technology's ethical development, application, and regulation. This theory creates a context where technology can advance society while upholding fundamental human values.

1.2.6   ETHICAL PLURALISM: BALANCING MULTIPLE THEORIES

Ethical pluralism—a philosophical stance recognizing multiple, potentially contradictory ethical principles—is particularly important in IT. As IT transforms society, various ethical issues, such as data privacy, AI, and digital access inequality, are emerging. Ethical pluralism provides IT professionals with a practical framework for navigating these issues by acknowledging the legitimacy of multiple ethical perspectives.

Ethical pluralism's utility in IT is demonstrated by its capacity to reconcile contradictory values, accommodate cultural peculiarities, and promote group problem-solving. It provides IT professionals a flexible arsenal for addressing evolving ethical issues and encourages multidisciplinary debates and stakeholder participation. Ethical pluralism can resolve ethical problems when one ethical theory falls short.

However, accepting ethical diversity is not without its challenges. Determining which ethical values should take precedence can take time and effort, which may lead to relativism or confusion. To establish a balance between universal ethical norms and diverse perspectives requires considerable thought. In conclusion, ethical pluralism equips IT professionals with the knowledge of making prudent decisions that promote responsible technology development, safeguard individual rights, and advance a morally virtuous digital society.

1.3 ETHICAL DECISION-MAKING MODELS FOR IT PROFESSIONALS

As technology continues to impact numerous facets of human existence and alter our environment, IT professionals frequently face complex ethical dilemmas requiring careful consideration and analysis. Several ethical decision-making models have been developed to resolve these issues and provide IT professionals with a formal framework for making well-informed and morally sound decisions [4, 5].

1.3.1 THE ETHICAL DECISION-MAKING PROCESS

The ethical decision-making process provides individuals, including IT professionals, with a methodical and reflective method for addressing complex ethical dilemmas. This procedure comprises a series of interconnected stages that assist individuals in analyzing, evaluating, and resolving ethical issues deliberately and principally (Figure 1.1).

1.3.1.1 Step 1: Ethical Awareness

The process begins with the recognition of an ethical conundrum. Identifying situations where conflicting values, interests, or principles create uncertainty or tension is required. IT professionals must develop the capacity to discern and acknowledge these ethical challenges, which frequently concern privacy, data security, intellectual property, and social impact.

1.3.1.2 Step 2: Gathering Information

Once an ethical issue has been identified, IT professionals must collect all pertinent information. This includes comprehending the situation's technical, legal, and social aspects. Effective decision-making requires a firm grasp of the relevant facts, prospective consequences, and the various stakeholders' respective interests.

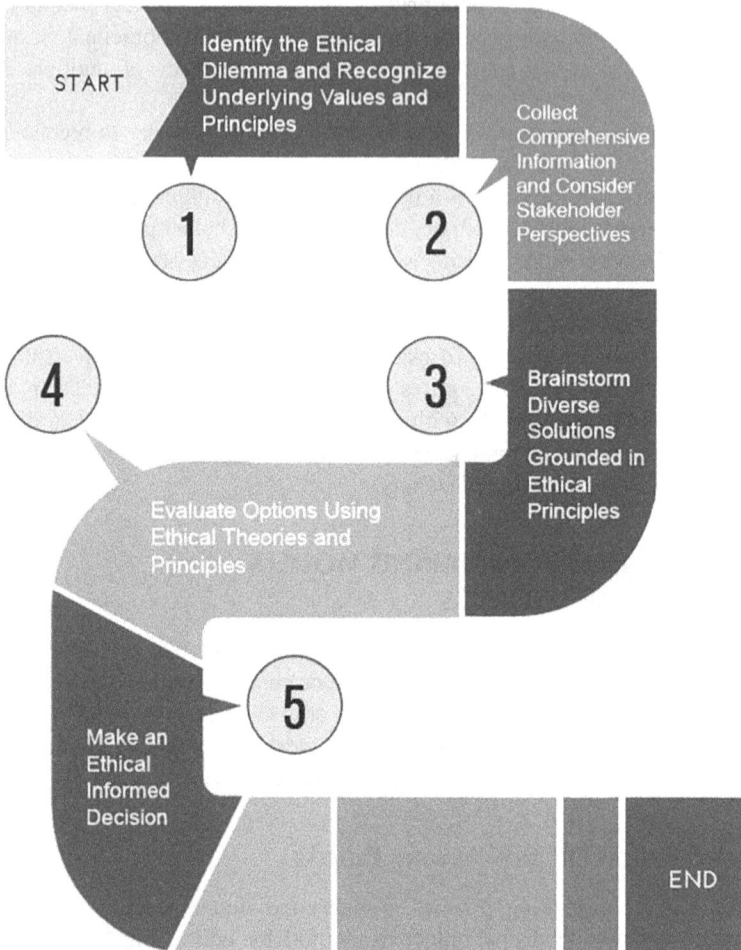

FIGURE 1.1 Ethical decision-making process for IT professionals

1.3.1.3 Step 3: Identifying Options

IT professionals generate various potential solutions to the ethical dilemma at this stage. It is essential to consider multiple options that adhere to ethical principles and professional standards. Creativity and critical thinking are crucial as IT professionals evaluate alternative future courses of action.

1.3.1.4 Step 4: Ethical Analysis

Each prospective option is evaluated based on well-established ethical theories or principles by IT professionals conducting an ethical analysis. This involves evaluating the options' congruence with such values as honesty, fairness, respect for individuals, and societal welfare. Ethical analysis also considers each alternative's potential advantages, disadvantages, and trade-offs.

1.3.1.5 Step 5: Decision-Making

After conducting an exhaustive ethical analysis, IT professionals conclude. They choose the alternative that appears to be the most morally justifiable and consistent with their professional responsibilities. This decision is reached by evaluating the various ethical factors and determining the course of action that maximizes ethical integrity and minimizes negative consequences.

1.3.1.6 Step 6: Implementation

After making a decision, IT professionals implement the selected course of action. This may entail communicating the decision to relevant stakeholders, developing a detailed plan, and taking the necessary actions to address the ethical dilemma effectively. Transparency and effective communication are fundamental elements of a successful implementation.

1.3.1.7 Step 7: Reflection and Learning

IT experts reflect and learn once a problem has been fixed. They assess the results of their choice and think about the wider ramifications of situations like this in the future. With the help of this reflection, they can hone their ethical reasoning and decision-making abilities and continuously improve how they handle ethical dilemmas.

1.3.2 PROMINENT ETHICAL DECISION-MAKING MODELS

Various ethical decision-making frameworks can substantially help IT workers negotiate the complicated terrain of their profession, empowering them to tackle difficult ethical issues with care and precision. These models offer systematic methods that enable IT professionals to make well-informed, morally sound decisions, upholding the highest standards of professionalism and integrity. The following are several well-known ethical decision-making frameworks that are especially pertinent to IT professionals.

The Potter Box Model. Rooted in a four-step process—definition, values, principles, and loyalties—the Potter Box Model systematically guides IT professionals to dissect ethical issues [6] methodically. By defining the problem, identifying underlying values, formulating ethical principles, and considering loyalties to various stakeholders, IT practitioners can systematically analyze complex situations and arrive at a balanced and ethically sound decision.

The SAD Formula. Standing for Software, Authority, and Duty, the SAD Formula offers a structured framework for IT professionals to assess their decisions' legal, authoritative, and duty-bound aspects. This model encourages a comprehensive evaluation of the decision's impact on software integrity, adherence to authority, and fulfillment of professional duties, promoting a holistic understanding of ethical implications.

The Nine-Box Model. Designed to facilitate a comprehensive analysis of ethical challenges, the Nine-Box Model urges IT professionals to consider various factors, including legality, public relations, and personal conscience. By evaluating decisions across multiple dimensions, IT practitioners gain a more nuanced perspective on the ethical landscape, enabling them to make well-balanced and principled choices.

The Wall Street Journal Model. Acknowledging the significance of public perception and stakeholder interests, the Wall Street Journal Model encourages IT professionals to contemplate how the public and stakeholders might interpret their decisions. By embracing transparency and anticipating potential reactions, IT practitioners can make decisions that align with ethical principles and contribute to a positive public image.

The Markkula Center Framework. Developed by the Markkula Center for Applied Ethics, this comprehensive framework offers a step-by-step approach that encompasses gathering information, identifying stakeholders, applying ethical principles, and assessing potential consequences. By systematically walking through these stages, IT professionals can ensure a well-rounded and thorough examination of ethical considerations in their decision-making process.

Adopting these ethical decision-making frameworks gives IT professionals useful tools to deal with the complex ethical issues that are part of their job. In order to ensure that technology improvements are pursued in a way that is consistent with moral standards and societal values, these models build a culture of responsible innovation, openness, and accountability. By incorporating these frameworks into their decision-making process, IT workers may achieve a healthy balance between pushing the limits of technological innovation and respecting the ethical norms that support a sustainable and ethically conscious technological landscape. In the end, the thoughtful implementation of these models leads to the favorable and long-lasting effects of IT on society at large.

1.4   THE IMPORTANCE OF ETHICAL LEADERSHIP IN IT ORGANIZATIONS

IT businesses' cultures and practices are significantly shaped by ethical leadership. The ethical issues relating to technology's creation, implementation, and use have grown more complicated and important as it continues to advance at an unparalleled rate. IT workers must use ethical leadership as a compass to traverse these challenges with honesty, accountability, and a dedication to societal well-being [5]. Table 1.1 provides a breakdown of different aspects of ethical leadership within IT businesses, explaining their roles, giving examples, and highlighting the benefits they bring.

1.4.1   SETTING THE ETHICAL TONE

Senior executives and managers in IT businesses should set an example for ethical conduct from the top down. They create the moral guidelines by which the company conducts its business, emphasizing principles like integrity, openness, privacy, respect, and a commitment to reducing harm. Leaders cultivate an environment where people value moral decisions by regularly exhibiting and supporting these ethical principles.

1.4.2   FOSTERING TRUST AND ACCOUNTABILITY

Ethical leaders in IT firms foster an atmosphere of trust and responsibility. They promote open dialogue and allow staff members to voice ethical issues without fear of retaliation. Through this open discussion, possible ethical problems are detected and resolved before they become more serious problems that could harm people or society.

TABLE 1.1
Role of ethical leadership in IT businesses

Aspect	Role of Ethical Leadership	Examples	Benefits
Setting the Ethical Tone	Establish moral guidelines; emphasize integrity, openness, and privacy; cultivate an ethical environment	Senior leaders modeling ethical behavior; emphasize honesty and respect	Builds trust and a positive work culture
Fostering Trust and Accountability	Create trust; encourage open dialogue; resolve ethical concerns	Encourage employees to voice concerns; address issues promptly	Enhances communication, prevents ethical lapses
Balancing Innovation and Responsibility	Balance advancement with ethics; consider consequences	Innovate responsibly; assess social and environmental impacts	Ensures ethical development of technology
Navigating Ethical Gray Areas	Address ambiguity; make thoughtful decisions; consult experts	Tackle uncertain situations; seek guidance when needed	Upholds ethical standards in complex scenarios
Mitigating Unintended Consequences	Proactively consider implications; value risk analysis; prioritize well-being	Anticipate effects of technology; minimize potential harm	Reduces negative societal and environmental impacts

1.4.3 BALANCING INNOVATION AND RESPONSIBILITY

The conflict between innovation and responsibility is one of the industry's particular difficulties. Ethical leaders find a balance between promoting technological advancement and making sure it adheres to moral standards. They lead their teams as they investigate cutting-edge solutions while considering any potential social, environmental, and ethical repercussions. This strategy guards against the heedless pursuit of advancement at the expense of moral issues.

1.4.4 NAVIGATING ETHICAL GRAY AREAS

Leaders frequently meet ethically ambiguous situations in the modern IT industry where more than established standards and guidelines may be required. When faced with uncertainty, ethical leadership requires making well-thought-out decisions, considering other viewpoints, and consulting experts if required. Leaders set an example for their teams to tackle challenging circumstances wisely and responsibly by committing to ethical discernment.

1.4.5 MITIGATING UNINTENDED CONSEQUENCES

Technology improvements can have significant harmful and beneficial effects. Ethical executives in IT firms proactively consider and mitigate the unexpected implications of their projects and products. They highly value thorough risk analyses and practice ethical foresight to foresee potential societal, economic, and environmental effects. This proactive strategy guards against harm and shows concern for the stakeholders' long-term welfare.

1.5   ADDRESSING ETHICAL CHALLENGES IN THE IT INDUSTRY

Ethical issues have merged into the fabric of the IT sector, which is dynamic and always changing. Various ethical conundrums are presented as technology continues to change and redefine how we live, work, and interact; these conundrums require serious analysis and prompt resolution.

1.5.1   DATA PRIVACY AND SECURITY

Data security and privacy have become crucial ethical issues in the field of IT in today's linked digital world. A strong framework is needed to safeguard sensitive data from unwanted access, breaches, and misuse due to the unprecedented flow of massive volumes of data across numerous platforms. Data privacy centers on people's rights to manage their personal information and to know how businesses gather, use, store, and distribute it.

This environment presents a variety of difficulties. High-profile data breaches have made personal and corporate information vulnerable, underscoring the urgent need for strict security measures. As data moves through third-party systems, worries about responsible treatment and protection grow. It has become challenging to balance the need for surveillance for security and the right to privacy for each person. Furthermore, the complicated interplay between international data flows and local rules highlights the difficulty of preserving data privacy in a connected world.

IT experts play a crucial role in addressing these issues. They diligently seek to strengthen security measures by adhering to concepts, including data minimization, encryption, and access control. Regular audits and open communication of privacy policies underscore the dedication to protecting user data. User empowerment is equally vital, as it gives people control over their data and the ability to decide how to utilize it. The ethical treatment of data becomes increasingly important as technology advances, becoming a sign of responsible IT practices and promoting trust and confidence in the digital world.

Data security and privacy lay the foundation for ethical considerations in the IT sector. Maintaining a careful balance between technical advancement and protecting individual rights is necessary to navigate this environment. IT professionals significantly contribute to a secure and reliable digital ecosystem by embracing these principles and adopting a proactive approach, supporting moral ideals while directing technology development.

1.5.2   AI AND AUTOMATION

Automation and AI have quickly developed into revolutionary forces changing industries, economies, and civilizations worldwide. These once-science-fiction-only technologies now play crucial roles in everyday life, influencing commerce, communication, and decision-making. Automation comprises technology automating monotonous chores, whereas AI includes technologies that mimic human intelligence. Together, they promote productivity and efficiency in a variety of industries, including manufacturing, healthcare, finance, and transportation.

AI-powered robots perform complex jobs, AI-powered diagnostic tools assist healthcare experts, and algorithmic trading in finance serves many industries. Organizations can use their human resources more effectively for strategic objectives by automating routine chores. However, the quick use of AI and automation has sparked moral debates. Worries about employment displacement, biased algorithms, and the ethics of AI in conflict highlight the necessity for responsible development and cooperative solutions.

In the age of AI, accountability and transparency are crucial. It is imperative to comprehend how AI makes decisions, especially in crucial fields like healthcare and criminal justice. The future of work is also touched by automation, which may lead to the abolition of some positions while generating new career prospects in data science and AI research. In the end, while AI and automation have enormous potential, it is crucial to confront their ethical implications and use them responsibly to have a future that benefits society as a whole [7].

1.5.3 SOCIAL IMPACT AND RESPONSIBILITY

As technology continues to influence almost every aspect of contemporary society, the idea of social effect and responsibility has gained substantial prominence in info IT. People, businesses, and communities are turning to technology companies and experts to use their influence responsibly and constructively impact the larger social environment as IT solutions become more integrated into our daily lives.

There are several important factors to consider when dealing with this responsibility. In order to ensure that marginalized populations have access to technology, education, and opportunity, IT is crucial in closing the digital divide. This might entail programs to spread gadgets, offer digital literacy training, and provide affordable internet access. Second, it is crucial to use ethical data procedures. Respecting user privacy and receiving informed consent, trustworthy IT professionals and businesses manage data transparently and securely. They work to produce inclusive solutions that serve users of all backgrounds and abilities because they understand the value of diversity and inclusion within their businesses and the technology they design.

The IT sector is ideally situated to promote environmental sustainability. Technology firms contribute to a cleaner future by implementing green IT practices, such as lowering energy use and embracing renewable energy sources. In parallel, they engage in social innovation, utilizing their knowledge to provide tech-driven answers to urgent societal problems. This might entail utilizing blockchain for transparent supply chain management or data analytics for tracking diseases. To ensure that new technologies enhance human capabilities without introducing bias or ethical concerns, IT professionals are tasked with building AI systems that are fair, transparent, and accountable.

1.5.4 INTELLECTUAL PROPERTY AND COPYRIGHT

The ideas of intellectual property (IP) and copyright are crucial in forming ethical behaviors and promoting innovation in the dynamic field of IT. A creator's legal rights to protect their innovations, designs, and creative works are collectively called IP. Software, music, and other original physical products are particularly protected by copyright, a subcategory of IP.

Copyright is now faced with new difficulties because it is so simple to reproduce and distribute digital content in the modern world. For restricted uses, such as criticism, education, and research, without prior authorization, the fair use concept permits copyrighted material. There has also been an increase in open-source software, governed by licenses encouraging cooperation and openness in software development.

However, worries about piracy and unlawful sharing have arisen due to the broad availability of digital information. A framework for handling online copyright infringement is provided by the Digital Millennium Copyright Act (DMCA) [8]. However, disparities in copyright legislation worldwide make it difficult to enforce rights everywhere.

The ethical challenge of balancing the need for open information flow and authors' rights persists. To create a vibrant digital ecosystem that respects IP and copyright while stimulating innovation and collaboration, a balance between these factors must be struck through constant communication, legal development, and technological solutions.

1.5.5 EVOLVING WORKFORCE DYNAMICS IN THE IT INDUSTRY

The workforce dynamics in the IT sector have seen a significant upheaval due to technical improvements and changing attitudes toward the workplace. Breaking down geographical borders and providing flexibility, remote and distributed work has emerged as a defining characteristic. However, it also presents communication and team cohesion issues. Although it raises questions about employment stability and perks, the advent of the gig economy and freelancing has given IT professionals the freedom to choose tasks that align with their expertise.

Cross-cultural collaboration has been facilitated by globalization, enhancing IT teams with a variety of perspectives yet necessitating skill in navigating various working methods and time zones. Another result of the industry's rapid evolution is the requirement for ongoing skill development among IT employees. Although this adaptability boosts productivity, it calls for careful management of the blending of work and personal life to lower the risk of burnout.

This dynamic atmosphere emphasizes the value of diversity and inclusion and encourages businesses to welcome people from diverse backgrounds and experiences to foster creativity. A more collaborative and agile approach is becoming more prevalent in leadership and management styles. Ultimately, the changing workforce dynamics in the IT sector necessitate a proactive and flexible strategy to fully realize the potential advantages and successfully solve the corresponding obstacles.

1.5.6 Emerging Technologies

It is a field constantly innovating, giving rise to a wide range of new technologies set to reshape various sectors and human experiences. These emerging innovations have enormous potential because they answer difficult problems and open up new horizons of possibility. However, integrating them also necessitates a complex strategy considering ethical, social, and pragmatic factors.

This revolution is led by AI and machine learning (AI/ML), which allow computers to learn from data and transform a variety of industries. Although quantum computing, based on the ideas of quantum physics, provides computational capability that has never been seen before, it poses ethical concerns regarding data security and encryption. The Internet of Things (IoT) connects objects for automation, but it also raises privacy and security issues that need to be addressed.

Though biotechnology and genetic engineering can produce marvels in medicine and agriculture, serious ethical questions remain. Eco-friendly practices are fueled by sustainable technologies, while augmented reality (AR) and virtual reality (VR) change experiences and raise questions about privacy and immersion. Blockchain guarantees safe data exchange, and nanotechnology can change matter at the atomic level, calling for responsible development. Collaboration between stakeholders is essential for maximizing benefits and reducing dangers from these advances. By embracing ethical engagement, the IT sector can traverse this transformational landscape for the greater benefit. Figure 1.2 depicts ethical challenge solutions: AI/ML bias detection, quantum-safe encryption, AR/VR privacy controls, and genetic engineering guidelines, driving responsible tech advancement.

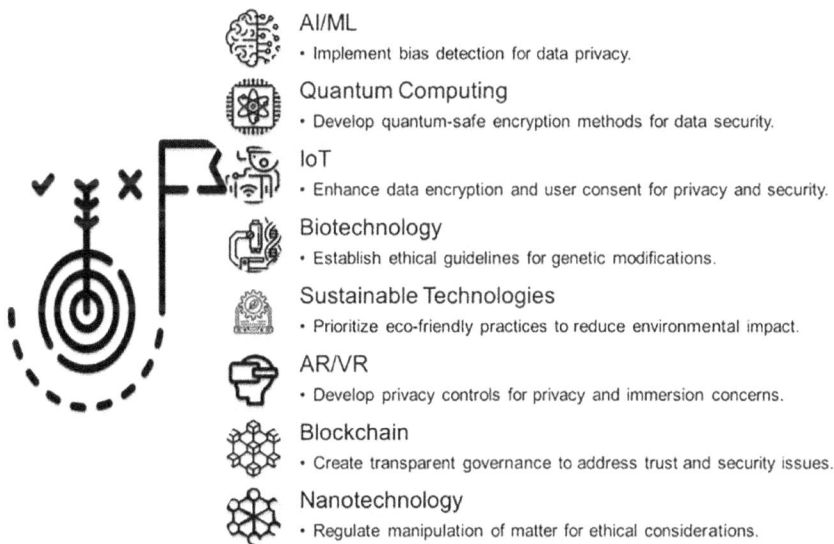

AI/ML
• Implement bias detection for data privacy.

Quantum Computing
• Develop quantum-safe encryption methods for data security.

IoT
• Enhance data encryption and user consent for privacy and security.

Biotechnology
• Establish ethical guidelines for genetic modifications.

Sustainable Technologies
• Prioritize eco-friendly practices to reduce environmental impact.

AR/VR
• Develop privacy controls for privacy and immersion concerns.

Blockchain
• Create transparent governance to address trust and security issues.

Nanotechnology
• Regulate manipulation of matter for ethical considerations.

FIGURE 1.2 Addressing ethical challenges in emerging technologies: strategies and solutions

1.6   THE ROLE OF ETHICAL CODES AND STANDARDS IN IT

Ethical norms and standards greatly influence IT professionals' conduct and decision-making. These regulations and standards offer a framework that encourages responsible and ethically upright behaviors in a field marked by quick technology breakthroughs and difficult ethical conundrums.

1.6.1   DEFINING ETHICAL CODES AND STANDARDS

The foundations of ethical behavior in the field of IT are ethical rules and standards. The Association for Computing Machinery (ACM) and the Institute of Electrical and Electronics Engineers (IEEE) both produced ethical codes, which are broad guidelines that summarize the beliefs and goals required of IT professionals. These codes offer aspirational direction and strongly emphasize moral qualities, including honesty, decency, fairness, and social responsibility. In contrast, ethical standards turn these general concepts into concrete rules that guide moral decision-making in particular circumstances. For instance, while a derived standard would outline procedures to secure user data, a code would emphasize data privacy.

In the IT industry, ethical norms and standards perform several essential tasks. They are a constant compass, guiding IT professionals through challenging moral conundrums. These standards help create a unified professional identity based on common ideals. They encourage accountability by providing a standard for moral conduct, increasing public and professional trust, and aiding conflict resolution by providing a framework for weighing competing interests. Additionally, they foster the industry's ongoing development and advancement through a culture of ethical awareness.

1.6.2   GUIDING PROFESSIONAL CONDUCT

An essential component of ethical norms and standards, guiding professional conduct in IT directs practitioners toward actions that uphold honesty, justice, and accountability. These norms cover various values influencing moral conduct and promoting a morally upright IT sector. Professionals are guided by honesty and integrity, stressing open communication and accepting responsibility for mistakes. Upholding each person's rights and dignity is crucial for fostering diversity and eliminating discrimination. Confidentiality, privacy, and data security must be maintained to ensure ethical information management. Professionals must also produce precise, high-quality work while creating trust by being upfront and transparent about technology's potential drawbacks and hazards. In order to minimize bias and distribute resources impartially, fairness and equity are emphasized.

Additionally, a strong sense of social responsibility compels IT workers to think about how their job will affect society and examine its ethical ramifications. This all-encompassing strategy promotes lifelong learning, ongoing professional development, and proactive adaptation to developing ethical quandaries. These guiding principles ensure that IT workers navigate the complicated world of technology with a firm commitment to moral behavior, which is advantageous to both their immediate stakeholders and society.

1.6.3 Balancing Technological Advancements and Ethical Considerations

The exponential growth of technology innovation offers both amazing benefits and complex challenges. While these inventions can potentially revolutionize society, they also spark ethical questions that demand serious thought. Finding a harmonious balance between ethical principles and technical advancement has become crucial in our linked society.

From conception to implementation, ethical considerations must be incorporated into technical design. This proactive strategy, often known as "ethics by design and default," guarantees that moral ideals govern all growth aspects. It highlights the importance of anticipating and managing potential dangers, encouraging goodness, and preventing harm. To holistically evaluate the ethical implications of emerging technologies, it is imperative that various stakeholders, including technologists, ethicists, policymakers, and cultural representatives, work together.

The diversity of ethical norms across cultures and the world must also be acknowledged to achieve this equilibrium. Respecting different viewpoints and values as technology crosses national boundaries is crucial. Collaborative and interdisciplinary debates ensure that discoveries adhere to widely accepted moral standards. We pave the way for a future where innovation is properly used to benefit society's well-being and growth by negotiating the complex interplay between technology advancements and ethical considerations.

1.6.4 Promoting Accountability and Trust in Information Technology

The development of responsibility and trust is a vital requirement in the dynamic world of IT. These guidelines, together with ethical concerns, form the cornerstone of ethical IT practice, affecting interactions with stakeholders and establishing the moral compass of the sector.

Building trust is based on two pillars: transparency and honesty. Users and clients are more likely to trust honest IT workers about their activities, decisions, and system operations. IT professionals build trust in an era characterized by cybersecurity concerns by prioritizing data privacy and security. A regular supply of trustworthy IT solutions develops gratifying connections and fosters a reputation for excellence, reliability, and quality, further reinforcing confidence.

Accountability and ethical decision-making go hand in hand, emphasizing making moral decisions even in difficult circumstances. Demonstrating a commitment to development and progress, admitting mistakes, and drawing lessons from setbacks improve the basis of trust. Additionally, encouraging community involvement and collaboration within the IT industry raises ethical standards, reaffirming the sector's dedication to ethical business practices.

Promoting IT responsibility and trust, in other words, goes beyond technical competence. It is an all-encompassing strategy where open communication, safe procedures, moral decision-making, and a culture of continual improvement come together to create a technological environment built on dependability and moral integrity. IT workers build the foundation for a time when technology and society work together for the greater good by following these principles.

1.6.5 Addressing Complex Ethical Dilemmas

Professionals working in the dynamic field of IT typically face complex ethical problems resulting from societal ramifications, technological advancement, and moral issues. These intricate problems need a methodical strategy incorporating moral consideration, teamwork, and a firm dedication to upholding integrity.

It takes ethical analysis and reasoning to get through these complex ethical conundrums. IT specialists must critically analyze the problem, pinpoint any underlying moral dilemmas, and weigh the pros and cons of various courses of action. Utilitarianism, deontology, and virtue ethics are just a few examples of ethical frameworks that provide helpful views for looking at moral quandaries from various viewpoints and assisting in investigating workable answers, even in the face of opposing values.

In order to solve complicated ethical problems, cooperation and stakeholder involvement are essential. Collaboration with a wide range of stakeholders, such as engineers, ethicists, legal professionals, and those who are directly impacted by the challenge, is required of IT professionals. This group feedback encourages thorough decision-making and reveals details that could go unnoticed. Furthermore, a forward-thinking mindset is necessary because professionals must think about how their decisions will affect society over the long term to ensure that they support sustainable growth and societal well-being.

IT professionals can successfully handle challenging ethical conundrums while following the industry's rules and standards by exercising ethical leadership, being open and honest in communication, and being dedicated to accountability. They support the responsible development of technology while defending the interests of individuals and society by fusing analytical rigor with empathy and a commitment to ethical ideals. Addressing difficult ethical conundrums continues to be a key component of ethical IT practice in a constantly changing technology environment. Table 1.2 summarizes the key aspects of addressing complex ethical dilemmas in the dynamic field of IT, highlighting their significance, actionable approaches, and outcomes.

TABLE 1.2
Strategies for addressing complex ethical dilemmas in IT

Factor	Explanation	Importance	Actionable Approach
Complex Ethical Problems in IT	Ethical challenges from tech progress, society, morals.	Guiding integrity in dynamic IT landscape.	Methodical strategy incorporating moral consideration, teamwork, and dedication to integrity.
Ethical Analysis and Reasoning	Analyzing, identifying dilemmas, and weighing options.	Diverse ethical frameworks aid informed decisions.	Critical evaluation, moral issue identification, and pros/cons assessment using ethical theories.
Cooperation and Stakeholder Involvement	Collaborating with diverse experts, forward-thinking.	Comprehensive solutions, long-term societal welfare.	Collaboration, stakeholder engagement, and forward-thinking mindset for sustainable outcomes.
Ethical Leadership and Accountability	Leading ethically, transparent communication, and accountability.	Responsible tech growth, societal safeguarding.	Ethical leadership, transparent communication, accountability for decision-making.

REFERENCES

1. Powell, A. B., Ustek-Spilda, F., Lehuedé, S., & Shklovski, I., Addressing ethical gaps in 'Technology for Good': Foregrounding care and capabilities. *Big Data & Society*, 2022. 9(2): p. 20539517221113774.
2. Marseille, E. and J.G. Kahn, Utilitarianism and the ethical foundations of cost-effectiveness analysis in resource allocation for global health. *Philosophy, Ethics, and Humanities in Medicine*, 2019. 14(1): pp. 1–7.
3. Megías, A., L. de Sousa, and F. Jiménez-Sánchez, Deontological and consequentialist ethics and attitudes towards corruption: A survey data analysis. *Social Indicators Research*, 2023. 170(2): pp. 507–541.
4. Suarez, V. D., Marya, V., Weiss, M. J., & Cox, D, Examination of ethical decision-making models across disciplines: Common elements and application to the field of behavior analysis. *Behavior Analysis in Practice*, 2023. 16(3): pp. 657–671.
5. Whittier, N.C., S. Williams, and T.C. Dewett, Evaluating ethical decision-making models: A review and application. *Society and Business Review*, 2006. 1(3): pp. 235–247.
6. Jones, D.A., Emphasis on ethical awareness: Why ethics? Why now. *Journal of the Medical Library Association: JMLA*, 2014. 102(4): p. 238.
7. Bertoncini, A.L.C. and M.C. Serafim, Ethical content in artificial intelligence systems: A demand explained in three critical points. *Frontiers in Psychology*, 2023. 14: p. 1074787.
8. Pryor-Darnell, T.A., N. Andersen, and S. Rowling, Professional ethics, copyright legislation and the case for collective copyright disobedience in libraries. *Journal of the Australian Library and Information Association*, 2019. 68(2): pp. 146–163.

2 Ethical Responsibilities for IT Workers and Users

2.1 PROFESSIONAL ETHICS FOR IT PRACTITIONERS

Ethical considerations are of the uttermost importance in the field of IT, where technological advancements and digital innovation have an impact on the contemporary world. Software developers, system administrators, and data analysts have significant ethical obligations as part of their jobs. These obligations extend beyond specialized knowledge to encompass the effects of their work on individuals, society, and the environment.

The ethical environment in which IT professionals operate is complex, drawing from various philosophies and ethical systems. Among the principles that guide moral judgment are utilitarianism, deontology, virtue ethics, and the social contract theory. By understanding and traversing these frameworks, IT professionals can evaluate the consequences of their decisions, assess their responsibilities, and uphold values consistent with the general welfare.

2.1.1 CODE OF ETHICS FOR IT PROFESSIONALS

Integrity is essential as the foundation of responsible behavior in the dynamic field of IT. As the architects of digital progress, IT professionals are held to a strict code of ethics that emphasizes their commitment to honesty, respect, and the broader societal impact of their work [1].

This code is committed to absolute integrity and transparency, fostering trust through clear communication and ethical conduct. IT professionals safeguard sensitive data while maintaining privacy and confidentiality. They actively pursue professional development and continually refine their skills to provide the highest levels of competence. Fairness and non-discrimination need to be practiced to promote inclusiveness and diversity as fundamental principles. A strong sense of social responsibility motivates IT professionals to create technology that enhances society while minimizing its negative effects. They manage the complexity of their creations with openness and responsibility, ensuring that any potential damage is mitigated and remedied. Figure 2.1 shows the ethical guidelines and key responsibilities that shape the behavior and decision-making of IT professionals.

 DOI: 10.1201/9781003584452-2

Promoting Digital Citizenship

Respect Others
Thoughtful Sharing
Protect Privacy

Safeguarding Privacy and Data Protection

Configure Privacy Setting
Obtain Consent and Permission
Avoid Deception

Addressing Misinformation and Fake News

Critical Evaluation
Verify Sources
Share Responsibility

Balancing Personal and Professional Identity

Transparency
Maintain Reputation
Disclose conflict of Interest

FIGURE 2.1   Code of ethics for IT professionals' flowchart

2.1.2 BALANCING TECHNICAL COMPETENCE AND ETHICAL CONSIDERATIONS

The expertise of IT professionals is a potent force propelling digital innovation in the world of IT, which is constantly evolving. These experts, who are knowledgeable in programming, system administration, data analysis, and other disciplines, can influence the digital environment. With this technological proficiency, however, comes the requirement to navigate innovation with a comprehensive analysis of its potential repercussions.

This equilibrium between technical expertise and ethical consideration requires a diversified strategy. Since every technical choice has far-reaching societal repercussions, IT professionals need to go beyond the confines of code and system design. To achieve this equilibrium, one needs to be proactive in anticipating potential problems and responsive to ethical issues. Using ethical foresight, professionals can make decisions that prioritize the welfare of individuals and society as a whole.

A collaborative environment that fosters synergy permits the fusion of technical expertise and ethical awareness to flourish. To develop a comprehensive understanding of the ethical landscape, IT professionals should pursue insights from various disciplines, such as ethics, law, and sociology. This interdisciplinary collaboration equips specialists with the tools they need to comprehend the effects of their technical solutions thoroughly. IT professionals can effortlessly combine technical innovation with ethical rigor by keeping abreast of new ethical standards and incorporating ethical reflection into the design and development process.

2.2   ETHICAL OBLIGATIONS IN IT PROJECT MANAGEMENT

Achieving success in IT project management requires more than just strategic planning, technical expertise, and strict adherence to ethical standards. Ethical considerations are essential for ensuring that IT initiatives are executed responsibly, transparently, and morally. IT project managers are responsible for upholding these ethical obligations throughout the project.

2.2.1   ENSURING TRANSPARENCY AND ACCOUNTABILITY

The foundational principles of ethical IT project management are transparency and accountability. These principles promote open communication, truthful reporting, and responsible decision-making throughout the project lifecycle. IT project managers are crucial in establishing and sustaining a culture of transparency and accountability, fostering stakeholder trust, and assuring the project's success.

2.2.1.1   Consistent and Honest Communication

Transparent communication is required to keep stakeholders abreast of the project's development, obstacles, and accomplishments. Project managers empower stakeholders to make informed decisions and effectively manage their expectations by providing precise and timely updates. By adhering to open and honest communication, project managers empower stakeholders to stay engaged, contribute insights, and collaborate toward successful project outcomes. Table 2.1 outlines key actions supporting this principle, their descriptions, and associated benefits.

TABLE 2.1
Effective communication strategies

Action	Description
Status Reporting	Regularly share project status reports detailing accomplishments, milestones, risks, and issues. Highlight both positive developments and challenges.
Open Dialogue	Encourage open discussions during team meetings, allowing team members to express concerns and share insights.
	Address questions and provide clarifications to ensure everyone has access to relevant information.
Transparent Documentation	Maintain easily accessible project documentation, including project plans, scope documents, and meeting minutes.
	This documentation should be clear and organized, enabling stakeholders to review project details when needed.

2.2.1.2   Accurate Reporting

Accurate reporting involves presenting project information truthfully and without bias. Providing honest and precise reports ensures that stakeholders clearly understand the project's financial, technical, and operational aspects. Table 2.2 outlines the key practices associated with accurate reporting in project management.

TABLE 2.2

Key practices for accurate project reporting

Key Practices	Description
Accurate Reporting	Accurate reporting involves presenting project information truthfully and without bias. Providing honest and precise reports ensures that stakeholders clearly understand the project's financial, technical, and operational aspects.
Financial Transparency	Maintain transparent financial records, outlining budget allocations, expenses, and any changes that impact project costs. Communicate how funds are utilized and any deviations from the original budget.
Risk Disclosure	Identify and communicate potential risks and uncertainties to stakeholders. Describe mitigation strategies and contingency plans, demonstrating a proactive approach to risk management.
Performance Metrics	Define and track performance metrics that align with project objectives. Present data objectively, highlighting achievements, areas for improvement, and any variations from projected outcomes.

2.2.1.3 Timely Issue Resolution

Addressing problems and conflicts promptly prevents them from escalating and negatively impacting the project's progress and reputation. Ethical project managers take swift action to resolve issues and maintain a positive project environment. The key practices outlined in Table 2.3 encompass effective project management strategies.

TABLE 2.3

Key practices for timely issue resolution

Key Practices	
Proactive Problem-Solving	Anticipate challenges and proactively address them to prevent disruptions. Encourage team members to report issues promptly and provide mechanisms for raising concerns.
Conflict Management	Establish a process for managing conflicts within the team and among stakeholders. Mediate disputes impartially, seeking solutions that prioritize the project's interests.
Escalation Procedures	Define escalation paths for issues that cannot be resolved at the project level. Ensure stakeholders know how to escalate concerns and access higher levels of management if necessary.

2.2.2 Handling Stakeholder Relationships and Conflicts of Interest

Maintaining reliable stakeholder relationships while avoiding potential conflicts of interest is essential for project success and maintaining ethical standards in the ever-changing IT project management landscape. The results of a project are substantially influenced by its stakeholders, which consist of individuals, groups, and organizations affected by or invested in it. Ethical project managers prioritize responsible decision-making and encourage cooperation and trust among stakeholders.

Ethical project managers are aware of their responsibility to serve all stakeholders equitably and impartially, regardless of their preferences, connections, or external pressures. They prioritize the success of the undertaking as a whole and the legitimate interests of stakeholders over any potential personal gain. The equitable distribution of resources demonstrates this commitment, the assurance that stakeholders receive accurate information, and the proactive resolution of any potential conflicts of interest.

Different agendas, objectives, or expectations of stakeholders may result in conflicts of interest. Ethical project managers are adept at detecting and resolving these issues by being transparent with one another, collaborating, and employing effective conflict-resolution techniques. Ethical project managers resolve disputes that benefit all parties by nurturing open and sincere communication among stakeholders, employing mediation procedures, and seeking win-win solutions.

Transparency is crucial to the management of stakeholder relationships and conflicts of interest. Transparency not only ensures accountability but also promotes stakeholders' trust and credibility. Ethical project managers recognize the importance of disclosing any personal, financial, or professional interests that could influence their decision-making or compromise the project's integrity. Project managers create an environment where stakeholders can confidently participate in the project process through full disclosure, rigorous documentation, and adherence to ethical standards.

2.2.3 Addressing Ethical Dilemmas in Project Decision-Making

During project decision-making in the dynamic field of IT, moral conundrums frequently manifest as complex problems. These dilemmas, which result from competing ideals, interests, and principles, require cautious resolution to preserve the integrity of project outcomes. To maintain unshakable moral integrity, ethical project managers utilize methodical methodologies and ethical decision-making frameworks, which are crucial for navigating these complex situations.

Models for making ethical decisions offer well-organized frameworks for systematically addressing ethical dilemmas. Deontology focuses on moral principles, whereas utilitarianism emphasizes maximizing general happiness. While ethical relativism acknowledges context-dependent differences, virtue ethics emphasizes moral qualities. Key steps in resolving ethical dilemmas include recognizing the dilemma, obtaining input from stakeholders and experts, analyzing options through an ethical lens, assessing consequences for all stakeholders, reaching a principled decision, transparently communicating the decision, and reflecting on results to improve future decision-making.

Ethical project managers employ a shrewd approach to address the complexities of ethical challenges. They collaborate to ensure that diverse perspectives are considered, that ethical standards are upheld and that the chosen course of action establishes a balance between immediate objectives and long-term ethical considerations. By efficiently implementing ethical frameworks and promoting open communication, these managers steer projects toward results that are technically sound and morally exemplary contributions to the organization and society at large.

2.3 ETHICAL IMPLICATIONS OF IT OUTSOURCING AND OFFSHORING

Businesses looking for cost reductions, access to specialized expertise, and enhanced flexibility frequently use IT outsourcing and offshoring. While these methods have some advantages, they also bring up serious ethical issues that IT professionals and companies must carefully negotiate. This section examines the main ethical ramifications of IT offshoring and outsourcing and provides advice on handling them.

2.3.1 Evaluating Social and Economic Impact

IT outsourcing and offshore activities significantly impact the social and economic landscapes of the home country of the outsourcing business and the host country. Organizations need to morally assess the potential social and economic repercussions of their decisions as they work to streamline their processes and cut costs [2].

2.3.1.1 Job Displacement and Economic Disparities

The possible displacement of local employees is one of the main ethical issues with IT outsourcing and offshoring. Ignorant migration to offshore sites on local labor markets and economies is impossible. Assessing potential implications on employment possibilities and contributing to measures that address job displacement are duties of ethical IT practitioners and organizations.

Questions about whether the organization is causing higher unemployment rates or economic inequality in the host nation arise when assessing the social impact. To prevent the advantages of outsourcing from harming vulnerable populations, balancing cost savings and assisting local communities is essential.

2.3.1.2 Ethical Considerations for Economic Growth

Organizations may benefit financially from IT outsourcing and offshoring, but ethical considerations need to go beyond short-term cost savings. The lens of sustainable economic growth should be used to examine these actions by responsible enterprises. This entails determining whether outsourcing adheres to the moral principles of shared prosperity and promotes the long-term economic development of the host nation.

Ethical IT practitioners should actively talk with stakeholders, including local communities, governmental organizations, and industry partners, to understand the possible financial repercussions of their outsourcing decisions. Organizations can make more informed decisions that are moral and in line with their social obligations by considering the broader economic repercussions.

2.3.1.3 Supporting Local Communities

Organizations can actively support the local communities impacted by outsourcing and offshore to lessen the negative social effects of these actions. This may entail supporting initiatives to create new jobs, encouraging skill development and educational programs for displaced employees, and working with neighborhood organizations to promote economic resilience.

Moral IT workers might look at ways to interact with the community in the host nation, fostering social progress through charitable endeavors, knowledge sharing, and capacity building. Organizations can increase their positive influence and strengthen their ethical commitment by actively contributing to the growth and development of local communities.

2.3.2 MITIGATING RISKS IN OUTSOURCING RELATIONSHIPS

There are many advantages to outsourcing IT services and offshoring technology projects, including having access to specialized expertise, cost savings, and increased flexibility. However, these advantages come with built-in dangers that may jeopardize the efficacy, security, and morality of the task being outsourced. To guarantee that the outsourced projects adhere to the organization's ethical standards and goals, it is crucial to mitigate these risks.

A thorough due diligence procedure is the first step toward effective risk reduction when choosing an outsourcing partner. IT professionals and businesses should thoroughly evaluate potential partners' skills, performance history, and commitment to ethical behavior. This entails assessing the partner's financial soundness, technological proficiency, data security protocols, and prior client interactions. A thorough selection procedure reduces the likelihood of ethical lapses by identifying partners who share the organization's values.

Outsourcing partnerships should be based on ethical principles. Organizations should develop detailed ethical standards that define what is expected regarding data security, labor laws, environmental sustainability, and other relevant areas. These rules provide a foundation for the outsourced labor and offer a framework for moral judgments throughout the project. Organizations set the standard for responsible and conscientious outsourcing processes by stating their ethical standards.

A crucial ethical issue in outsourcing contracts is data security. Throughout the outsourcing process, IT professionals need to make sure that sensitive data and private information are protected. Strict data security requirements should be spelled out in contracts and access restrictions and compliance with data protection laws. Regular audits and evaluations can confirm that the outsourcing partner upholds the necessary data security and privacy standards.

Once an outsourcing connection has been created, constant oversight is necessary to guarantee that legal and ethical requirements are met. Organizations can monitor progress, spot potential ethical lapses, and resolve any departures from the agreed-upon terms by conducting regular performance evaluations. Transparency is essential; both parties should use open communication and reporting channels to promote reciprocal accountability and uphold moral consistency.

Having a clear mechanism in place for identifying and resolving ethical concerns or breaches is critical. Data breaches, infractions of labor laws, and environmental law violations are all examples of ethical violations. Organizations should lay out a clear escalation channel and corrective action plan to quickly resolve breaches. A timely response indicates a commitment to moral principles and assists in preventing the growth of ethical issues.

2.3.3 Respecting Cultural and Legal Differences in Offshoring

Offshoring IT activities and working with international partners are typical in today's connected global economy. These agreements, however, frequently include interactions between various cultural environments and judicial systems. IT professionals and companies involved in offshore activities have a crucial ethical duty to respect and manage these disparities.

Cultural awareness is a crucial component of ethical offshore. IT professionals need to be aware that other cultures have different communication norms, beliefs, and conventions. Understanding and adjusting to the cultural differences of offshore partners is necessary for effective communication and collaboration. This calls for attentive listening, empathy, and a dedication to mutual understanding to prevent misunderstandings and disputes.

Working under diverse legal frameworks while offshoring IT activities requires awareness of regional legal differences. Ethical practitioners require compliance with regional laws, rules, and industry standards. This includes abiding by labor rules, the environment, data protection, and intellectual property rights. Neglecting or avoiding these legal commitments may result in moral lapses and damage one's reputation.

One ethical challenge with offshore is ensuring that the personnel participating in IT operations is treated fairly and follows legal labor laws. To prevent offshore workers from being exploited or receiving unjust treatment, IT professionals should enquire about the working circumstances, pay, and perks. Promoting moral and socially responsible outsourcing requires cooperation with partners who support fair labor practices.

Cultural differences in ethical standards and values might be considerable. IT professionals and organizations need to be aware of any ethical differences between their standards and those of their offshore partners. Approaching ethical relativism, the notion that moral principles are influenced by culture, should be done with caution. Respecting cultural variety and supporting universal ethical standards like human rights and environmental sustainability need to be balanced.

IT personnel should participate in cross-cultural education and training to handle cultural and legal differences successfully. This entails becoming familiar with the regional laws, customs, and traditions. A smoother collaboration can be facilitated by educating offshore partners about one's own cultural and legal background. These initiatives foster respect for one another, improve communication, and assist avoid misunderstandings.

2.3.4 Balancing Cost Savings with Ethical Considerations

While cutting costs and increasing efficiency are respectable company objectives, balancing monetary gains and moral considerations carefully is critical. Maintaining this balance ensures that businesses and IT professionals are committed to ethical and responsible behavior.

2.3.4.1 Ethical Dilemma: the Cost-Effectiveness Trap

Due to the appeal of cost reductions through outsourcing or offshore, businesses may ignore or dismiss ethical issues. Decision-makers may be tempted to compromise on labor norms, environmental effects, and data security under financial pressure to cut costs. Due to this, businesses face a moral conundrum as they attempt to maximize advantages while minimizing harm.

2.3.4.2   Investing in Local Communities

A comprehensive analysis of the effects of outsourcing and offshore decisions is required to balance ethical issues and cost savings. One strategy is to invest in the health of regional communities impacted by these behaviors. Organizations can lessen the negative effects of job displacement and support sustainable economic development by offering training, education, and employment opportunities.

2.3.4.3   Ensuring Fair Wages and Working Conditions

Ensuring IT outsourcing and offshore workers receive fair pay, secure working conditions, and adequate benefits is a key component of ethical IT outsourcing and offshoring. Organizations should work with outsourcing partners to create and implement labor standards that promote human rights and dignity. By accomplishing this, IT professionals help create a more just and equal global workforce.

2.3.4.4   Long-Term Value over Short-Term Gains

While immediate financial gains may result from cost savings, ethical IT professionals understand the benefit of taking the long view. Rapid judgments made to save money might result in poor quality, data breaches, and reputational harm. Prioritizing moral factors like data security, quality control, and social responsibility can result in long-term success and a positive perception of your business.

2.3.4.5   Holistic Risk Assessment

A thorough risk assessment approach is necessary to balance ethical concerns with economic savings. Organizations should consider ethical, legal, reputational, and other potential risks while identifying and evaluating outsourcing and offshore options. To handle these risks, mitigation plans should be created, making sure that the quest for cost savings does not jeopardize morality.

2.3.5   PROMOTING TRANSPARENCY AND ACCOUNTABILITY

Open communication is the cornerstone of successful collaboration, which encourages trust between all participants. Establishing thorough communication channels ensures everyone is on the same page about the project's goals, schedule, and ethical considerations. Data security, privacy, ethical workplace conduct, environmental stewardship, and legal compliance require articulating ethical principles and guidelines. These concepts should be effortlessly included in agreements to emphasize the importance of ethical behavior.

Regular reporting and auditing procedures are essential to maintain transparency and hold stakeholders accountable. Regularly scheduled performance reviews, progress reports, and evaluations of ethical compliance guarantee uniformity in ethical obligations across the partnership. A whistleblower protection program strengthens openness by encouraging the disclosure of ethical problems without fear of retaliation. Processes for resolving conflicts need to be specified to enable timely and equitable resolution of ethical conundrums, highlighting ethical commitment even in difficult circumstances.

An attitude of constant improvement supports ethical behavior. A commitment to responsible outsourcing and offshoring is demonstrated through learning from mistakes, improving, and adjusting moral principles to changing conditions. This proactive strategy fosters a culture of honesty, fosters trust, and benefits the global IT landscape.

2.4   USER RESPONSIBILITIES AND ETHICAL USE OF TECHNOLOGY

As technology consumers, we are beneficiaries of its conveniences and stewards of its ethical application. Each individual is fundamentally obligated to use technology responsibly and ethically to foster a harmonious and just digital society.

2.4.1   Promoting Digital Citizenship and Responsible Technology Use

Digital citizenship has emerged as a guiding light in the ever-changing landscape of the digital era, illuminating the path to responsible and ethical technology use. This contemporary philosophy emphasizes the seamless incorporation of virtuous behavior from the physical world to the digital domain. Fundamentally, digital citizenship is a call to uphold the values of respect, empathy, and responsibility, thereby nurturing a harmonious online community.

Understanding the profound impact our actions have in the virtual realm is fundamental to the use of technology responsibly. Citizenship in the digital age entails navigating this landscape with intention, acknowledging the permanence of online content, and fostering a culture of empathy and positive interaction. It compels us to take a position against cyberbullying and harassment, protecting the Internet as an inclusive and safe haven.

The scope of responsible technology use includes intellectual property and information sharing. Digital citizenship entails upholding copyright laws, attributing sources, and fostering an environment of creativity and collaboration. In addition, this concept encompasses a broader mission: educating and empowering individuals of all ages to navigate the digital landscape with discernment. By teaching future generations critical thinking and media literacy, we equip them to make informed decisions and immunize them against the dangers of misinformation.

Digital citizenship is a collective effort that transcends individual displays. By adhering to its principles, we harmonize the digital world with the values we hold dear in our physical lives, creating a digital landscape that reflects the best of humanity: respectful, empathetic, and united.

2.4.2   Safeguarding Privacy and Data Protection

In a technologically advanced and interconnected world, preserving privacy and personal data has become a paramount ethical imperative. Users are responsible for upholding these principles when interacting with digital platforms, applications, and services to preserve their autonomy and the broader integrity of the digital landscape.

Data privacy encompasses controlling and administrating an individual's personal information, ensuring that it remains confidential and is used only by their desires. The digital realm flourishes on data frequently collected to enhance user experiences, deliver targeted content, and improve services. This data-driven ecosystem raises ethical concerns regarding the potential misuse, illicit access, and exploitation of personal data.

Users play a crucial role in shaping the landscape of data privacy. Users can control how much their data is shared and used by actively configuring their privacy settings. They need to maintain vigilance against intrusive data collection practices and exercise caution when granting permissions to applications and services. Additionally, advocating for comprehensive data protection laws and regulations helps establish a legal framework that protects user privacy.

The responsibility for data protection is not limited to individual users. Organizations and businesses that manage user data have a fiduciary responsibility to protect its security and confidentiality. Implementing robust cybersecurity measures, encryption protocols, and routine security assessments are essential safeguards for user data against breaches and unauthorized access.

User education and awareness are equally important components of data protection. Educating users on potential threats, such as phishing and social engineering, enables them to make informed decisions and implement practices that reduce vulnerabilities. By fostering a security-conscious culture, users contribute to an environment where the collective obligation for data protection is upheld.

Ethical quandaries frequently arise when weighing the benefits of data sharing against the potential for misuse. Users need to critically analyze the purposes for which their data is being collected and disseminated, ensuring that these purposes are consistent with their values and expectations. Transparent consent mechanisms and straightforward organizational communication enable users to make informed data-sharing decisions.

The ethical implications of data sharing transcend individual boundaries. Collective data sets have the potential for groundbreaking research and innovation, but responsible data-sharing practices need to be adhered to prevent unintended injury. Users, organizations, and policymakers should collaborate to establish guidelines for responsible data sharing that balance the pursuit of progress and the need to safeguard individual privacy.

2.4.3 Ethical Challenges of Social Media and Online Behavior

In the ever-changing realm of social media and digital communication, ethical considerations are of the utmost importance and shape the contours of online interactions and relationships. These platforms provide unprecedented avenues for expression and connection but present a complex array of ethical challenges that necessitate mindful navigation and responsible participation. Figure 2.2 illustrates the essential steps and considerations for ethically navigating social media and online platforms.

Authenticity and transparency in self-presentation are among the most important ethical considerations. Crafting polished online personas is common, but establishing a balance between showcasing highlights and maintaining authenticity is essential for fostering genuine connections and trust within online communities. Maintaining a level of openness ensures that the online realm reflects the realities of the offline world, fostering more meaningful interactions.

A further pressing ethical concern is the prevalence of misinformation and disinformation. The rapid dissemination of information on social media necessitates a greater obligation to fact-check and verify before sharing. By adhering to the principles of accuracy and veracity in one's sharing practices, one counteracts the proliferation of false narratives and contributes to a more informed and trustworthy digital discourse.

The rise of cyberbullying and online harassment highlights the importance of cultivating a respectful and compassionate online environment. Ethical users actively oppose and report detrimental behavior, thereby creating an online environment devoid of hostility. In addition, privacy and digital footprint management emerge as crucial ethical concerns. Users need to be mindful of the information they share, the permissions they grant, and the potential ramifications of their online presence. Effectively employing privacy settings and advocating for comprehensive data protection policies are essential for ethical digital engagement.

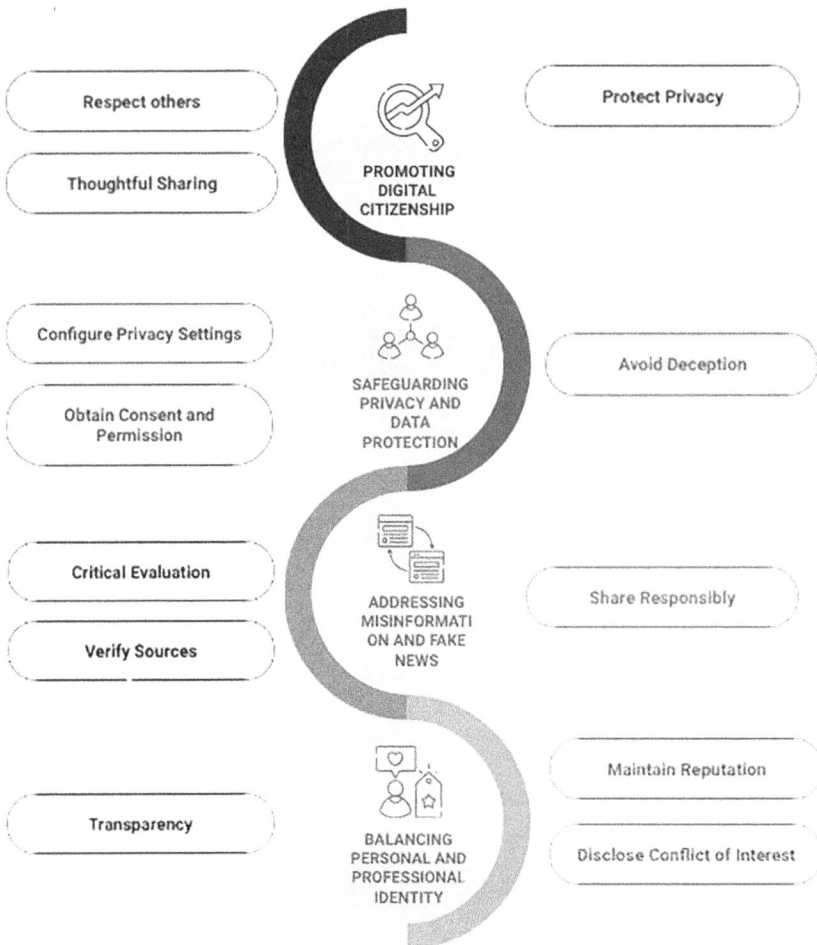

FIGURE 2.2 Managing ethical obstacles in social media and online conduct

2.5   ETHICAL CHALLENGES IN IT WORK ENVIRONMENTS

In the dynamic and ever-changing landscape of IT, ethical challenges in the work-place are pervasive and significant. As IT professionals collaborate on projects, interact with colleagues and stakeholders, and make important decisions, they face various ethical considerations that influence the industry's ethical fabric. It is crucial to address these obstacles to create an inclusive, respectful, and responsible IT work environment.

2.5.1   Fostering Inclusive and Diverse IT Workplaces

Fostering inclusive and diverse workplaces is not merely an ethical principle in the dynamic domain of IT; it is a strategic imperative that brings numerous benefits to individuals and organizations. A commitment to diversity and inclusion not only improves an organization's reputation but also fuels innovation, creativity, and over-all business success. Recognizing the value of diverse perspectives, experiences, and backgrounds, IT leaders and professionals need to advocate efforts to foster a thriving environment for all.

IT workplaces should actively seek out candidates from underrepresented groups and provide equal opportunities for development and advancement to embrace diversity as a driver of innovation. A diverse IT workforce begins with a reevaluation of employment and promotion practices. Blind recruitment practices, standardized evaluations, and inclusive job descriptions can assist in mitigating bias and leveling the playing field. Moreover, mentorship programs and leadership development initiatives enable individuals from diverse backgrounds to assume leadership positions and contribute to strategic decision-making.

The cornerstone of a diverse workplace is an inclusive corporate culture. IT organizations should foster a culture where all employees feel valued and respected and can voice their opinions. This involves establishing open commu-nication channels, actively listening to employees, and promptly addressing any instances of exclusion or bias. By encouraging dialogue about diversity, conduct-ing seminars on cultural awareness, and celebrating various cultural events, it is possible to break down barriers and create a cohesive, harmonious workplace. Recognizing employees' diverse requirements and responsibilities outside of the workplace and promoting work-life balance and flexibility expands the scope of inclusion.

2.5.2   Combating Discrimination and Harassment in the IT Industry

Discrimination and harassment continue to shadow the IT industry, impeding its development and potential. Whether anchored in gender bias, racial prejudice, or ageism, these harmful practices create an environment where individuals are mar-ginalized and their contributions are diminished. Recognizing the complexity of these issues is crucial, as is recognizing that discrimination and harassment can manifest in a variety of subtle and overt ways.

The IT industry needs to take a proactive stance against discrimination and harassment to combat this. The establishment of comprehensive anti-discrimination and anti-harassment policies is central to this effort. These policies should leave no room for ambiguity, defining what constitutes such conduct and highlighting the organization's commitment to nurturing a safe, respectful, and inclusive workplace. Regular training sessions and seminars should equip employees with the knowledge and tools to recognize, report, and prevent discrimination and harassment incidents.

It is essential to encourage reporting to break the cycle of discrimination and harassment. Those who choose to come forward need access to multiple reporting channels that guarantee anonymity and protect against retaliation. Equally essential is the availability of support mechanisms, such as counseling services, to assist victims in coping with the emotional aftermath of such experiences. Organizations can take effective action by enabling victims and witnesses to speak out.

Collaboration is crucial to the success of this group effort. The IT industry should collaborate with industry associations, advocacy organizations, and initiatives committed to diversity, equity, and inclusion. Stakeholders can amplify their efforts and promote systemic change by collaborating. The IT industry can break down the barriers of discrimination and harassment and pave the way for a more just and ethical future by nurturing a culture that holds perpetrators accountable, embraces diversity, and offers unwavering support to those affected.

2.5.3 WHISTLEBLOWING AND ETHICAL REPORTING OF WRONGDOING

In IT, where the rapid pace of innovation frequently intersects with complex ethical considerations, whistleblowing is a crucial mechanism for maintaining organizational transparency, accountability, and ethical integrity. Whistleblowing is the act of a person, typically an employee, informing the appropriate authorities or the public about unethical, unlawful, or harmful activities within their organization. This practice protects stakeholders' interests and plays a crucial role in preserving the IT industry's ethical fabric.

IT is not immune to ethical dilemmas and the possibility of misconduct. These may include data breaches and cybersecurity vulnerabilities, improper management of user data, and the development and deployment of technologies with negative outcomes. Whistleblowing enables conscientious IT professionals to speak out against unethical behavior without fear of retaliation, ensuring that ethical concerns are addressed and resolved appropriately. Whistleblowers prevent potential harm and steer organizations back onto ethical paths by revealing concealed or suppressed information.

To cultivate a culture that encourages ethical reporting and whistleblowing, organizations in the IT sector need to establish clear and accessible reporting channels. Whistleblowers should be afforded protection against retaliation so that they can come forward without fear of negative career consequences. This protection may include legal safeguards, assurances of confidentiality, and support mechanisms. The leadership should convey the significance of whistleblowing and emphasize its role in upholding the organization's ethical standards.

2.5.3.1 Ethical Reporting Process

Table 2.4 outlines a comprehensive framework for establishing an effective, ethical reporting process within organizations. Each step fosters a culture of integrity, accountability, and trust. Accessible channels, including anonymous options, provide diverse avenues for whistleblowers to come forward, while a timely response ensures swift and thorough actions upon receiving reports. Confidentiality safeguards sensitive information, protecting whistleblowers from potential retaliation, and fair treatment prevents discrimination or harassment. Holding wrongdoers accountable reinforces ethical conduct and signifies the organization's dedication to upholding its values. Despite these crucial steps, challenges such as resource limitations and navigating legal obstacles need to be addressed to ensure the successful implementation of this ethical reporting framework.

TABLE 2.4
Key steps for establishing an effective ethical reporting process

Step	Description	Challenges	Examples
1. Accessible Channels	Provide various reporting options, including anonymous avenues, to cater to different whistleblower preferences	Balancing anonymity with the credibility of reports	Ensuring proper verification of anonymous reports
2. Timely Response	Ensure swift and thorough action upon receiving reports, including comprehensive investigations and appropriate measures	Overcoming resource constraints for investigations	Limited personnel or expertise for complex cases
3. Confidentiality	Safeguard whistleblower identities and provide information, protecting them against potential retaliation	Safely storing and transmitting sensitive data	Implementing strong encryption methods
4. Fair Treatment	Guarantee fair treatment for whistleblowers, preventing retaliation, discrimination, or harassment against them	Identifying and addressing subtle forms of retaliation	Addressing microaggressions or subtle threats
5. Accountability	Hold wrongdoers accountable, showcasing the organization's dedication to upholding ethical standards at all levels	Navigating legal and internal resistance to change	Overcoming resistance from influential individuals

2.5.3.2 Collaborative Resolution

Rather than a confrontational act, whistleblowing should be viewed as a collaborative endeavor to uphold ethical standards. IT organizations should promote open communication between whistleblowers and management to facilitate the resolution of issues through constructive dialogue. This strategy promotes a culture of continuous development in which lessons learned from reported incidents contribute to refining ethical practices and preventing future issues.

2.6 ETHICAL ISSUES IN IT TRAINING AND EDUCATION

Education in IT significantly impacts the future of IT professionals and the ethical standards they uphold throughout their careers. To ensure the responsible and ethical development of the next generation of IT professionals, educators and trainers need to address a variety of ethical considerations.

2.6.1 ETHICAL CONSIDERATIONS IN CURRICULUM DESIGN AND DELIVERY

The ever-changing nature of IT necessitates that educators not only impart technical knowledge but also cultivate a profound appreciation for ethical implications. As curriculum designers for future IT professionals, we are responsible for fostering an ethical mindset.

The incorporation of ethical frameworks is fundamental to ethical IT education. By introducing students to theories like consequentialism, deontology, and virtue ethics, educators equip them to evaluate moral dilemmas methodically. It is essential to achieve a balance between technical expertise and ethical considerations. A cohort of IT professionals capable of resolving technical challenges and appreciating the ethical dimensions of their solutions can be fostered by combining technical expertise with an understanding of technology's broader impact.

Anticipating and resolving emergent ethical challenges is a further essential component. The rate of technological advancement generates new ethical dilemmas, necessitating curriculum modifications. Proactively incorporating concepts such as artificial intelligence ethics, algorithmic bias, and responsible data management is necessary. Fostering ethical reflection and dialogue is likewise essential. Creating an atmosphere where students engage in discussions, projects, and assignments that promote ethical reflection cultivates critical thinking and effective communication skills.

Collaboration with ethical experts and professionals outside of IT enhances the learning experience. Expert perspectives are provided through guest lectures, seminars, and partnerships with ethics, law, and social responsibility professionals. By emphasizing interdisciplinary collaboration, educators provide students with a comprehensive analysis of IT ethics. Collectively, these initiatives lay the groundwork for future IT professionals to approach their work with an ethical mindset, thereby ensuring the industry's positive impact on individuals and society.

2.6.2 Addressing Plagiarism and Cheating in IT Education

Academic integrity is of paramount importance in all educational contexts, including the field of IT. In IT education, honesty, originality, and responsible scholarship are essential for individual development and cultivating a culture of professionalism and ethical conduct within the industry. Plagiarism and cheating need to be addressed through a multifaceted strategy that includes explicit policies, proactive prevention, and educational interventions (Figure 2.3).

2.6.2.1 Plagiarism and Cheating

Plagiarism involves presenting someone else's work, ideas, or intellectual property as one's own without proper attribution. Conversely, cheating encompasses actions such as copying answers, using unauthorized materials during assessments, or collaborating without permission. Both practices erode the educational experience and undermine the development of essential skills vital for IT professionals.

2.6.2.2 Establishing Clear Expectations

Educators play a central role in preventing plagiarism and cheating by clearly communicating expectations and consequences. At the beginning of each course, instructors should outline their institution's policies on academic integrity, plagiarism, and cheating. This includes explaining what constitutes plagiarism, specifying appropriate citation methods, and defining acceptable collaboration practices. By setting the groundwork for ethical conduct, educators provide students with a clear understanding of the standards they are expected to uphold.

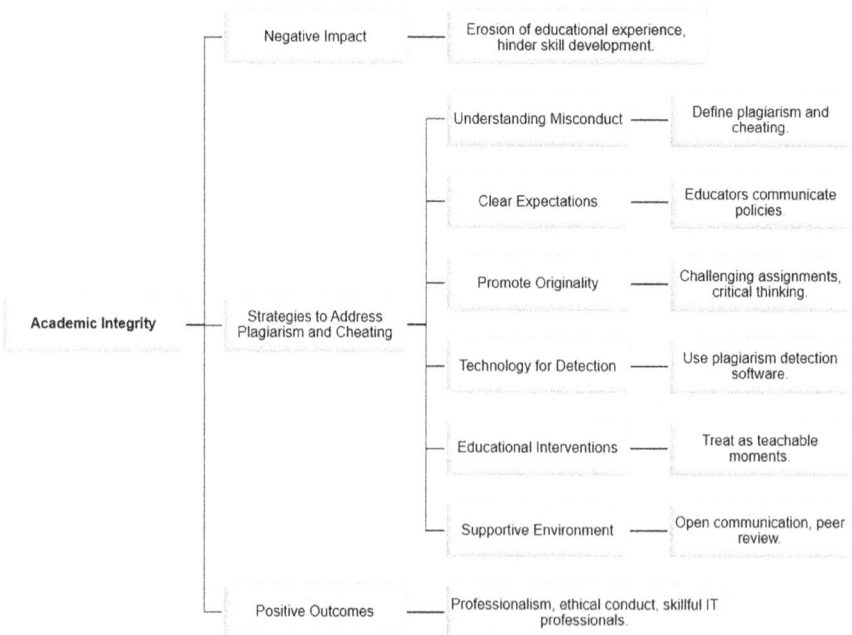

FIGURE 2.3   Strategies for addressing plagiarism and cheating in IT education

2.6.2.3   Promoting a Culture of Originality

Educators can encourage a culture of originality by designing assignments that challenge students to think critically and creatively. Projects that require problem-solving, analysis, and synthesis promote unique contributions and discourage reliance on external sources. Encouraging students to explore their perspectives and ideas fosters an environment where individuality is celebrated and rewarded.

2.6.2.4   Utilizing Technology for Detection

Technological tools can aid educators in detecting plagiarism and cheating. Plagiarism detection software can compare submitted work against a vast database of sources, highlighting potential instances of copied content. While technology can be a valuable aid, it should be used with a broader educational approach that emphasizes understanding and avoiding plagiarism rather than solely relying on punitive measures.

2.6.2.5   Educational Interventions and Remedies

When plagiarism or cheating is identified, educators should approach them as teachable moments. Instead of immediately resorting to punitive actions, instructors can engage students in discussions about academic integrity and its importance in the IT industry. Offering guidance on proper citation practices and providing resources for improving study habits can help students learn from their mistakes and make ethical choices in the future.

2.6.2.6   Fostering a Supportive Learning Environment

Creating an environment where students feel comfortable seeking help and guidance is crucial in preventing plagiarism and cheating. By establishing open lines of communication, educators can address any confusion students may have about citation, collaboration, or the expectations for assignments. Providing opportunities for peer review and group discussions also encourages cooperative learning while emphasizing the value of individual effort.

2.6.3   Encouraging Lifelong Learning and Professional Development

Education is not a one-time endeavor within the ever-changing IT landscape. Continuous professional development and lifelong learning are essential for a successful IT vocation. Educators play a crucial role in instilling a culture of lifelong learning in their students, allowing them to navigate a constantly evolving industry while confronting ethical challenges.

The significance of lifelong learning resides in recognizing the dynamic nature of IT. Educators emphasize that formal education is only the beginning, fostering an adaptable and inquisitive mindset. This mindset motivates people to search for new information, participate in industry forums, and embrace self-directed learning. By cultivating this mentality, students can remain proactive and adaptable in the face of technological changes [3].

Ethics are intertwined with lifelong education. Emerging technologies frequently present unanticipated moral dilemmas. Educators encourage students to critically analyze these repercussions critically, thereby nurturing their capacity to evaluate societal impacts and make ethically sound decisions. Participating in ethical dilemma-related discussions and case studies equips students to consider broader consequences and act ethically in their IT responsibilities.

Educational institutions promote lifelong learning by providing resources and guidance. Access to contemporary materials, online platforms, and industry events enables students to educate themselves independently. Educators act as mentors, guiding students and introducing them to professional opportunities. This supportive environment cultivates well-rounded, ethically conscientious IT professionals, allowing them to respond to industry changes and positively impact society by making informed decisions.

REFERENCES

1. Guerrero-Dib, J.G., L. Portales, and Y. Heredia-Escorza, Impact of academic integrity on workplace ethical behaviour. *International Journal for Educational Integrity*, 2020. 16(1): pp. 1–18.
2. Augier, M. and D.J. Teece, *The Palgrave encyclopedia of strategic management*. 2018: Palgrave Macmillan London.
3. Laal, M., Lifelong learning: What does it mean? *Procedia-Social and Behavioral Sciences*, 2011. 28: pp. 470–474.

3 Computer and Internet Crime
Ethical Implications and Mitigation

3.1 CYBERCRIME AND ITS IMPACT ON SOCIETY

Unprecedented connectedness and convenience have been made possible by the digital era, but it has also given birth to a brand-new kind of criminal behavior known as cybercrime. Cybercrime is the term for illegal acts that target people, companies, and countries using digital technology and the internet. This section digs into the complex world of cybercrime, looking at its different manifestations, moral ramifications, and prevention techniques.

3.1.1 THE EVOLUTION OF CYBERCRIME

Cybercrime, or the illicit use of digital technology to commit crimes, has seen a remarkable transformation from the early days of the internet (Figure 3.1). Cybercriminals modified their tactics as technology improved and society became more reliant on networked systems, taking advantage of new weaknesses to provide a complex, dangerous environment.

The 1970s and 1980s, when the internet was only starting, was when cybercrime first emerged. Technically skilled people took part in activities that would later provide the basis for contemporary cybercrime at this time. These activities included exchanging pirated software, breaking into computer systems without authorization, and experimenting with viruses and worms. Cybercrime during this time was often motivated by curiosity and a desire to test the limits of digital technology.

In the 1990s, as personal computers grew in popularity, hackers started using harmful software (malware) to take advantage of flaws in operating systems and software programs. Computer viruses were more prevalent during this period and propagated through infected files and email attachments, corrupting data and harming computers. Financial motivation also played a part, as some hackers started requesting ransom payments to decrypt encrypted information or systems.

In the late 1990s and early 2000s, as e-commerce expanded, there was an increase in financial cybercrime. Cybercriminals have changed their attention to collecting confidential financial data, including credit card numbers and login information for online banking. This led to internet fraud, phishing schemes, and identity theft. The introduction of digital currencies and online payment methods increased the potential for financial crimes.

DOI: 10.1201/9781003584452-3

Highly sophisticated cyberattacks conducted by nation-state actors and well-funded criminal organizations first appeared in the 21st century. Advanced persistent threats (APTs), which include persistent and deliberate attempts to penetrate systems and steal valuable intellectual property, have emerged as a key tactic. The widespread use of cyber espionage, political manipulation, and massive data breaches during this period illustrated the significant influence that cybercrime might have on world politics.

Cybercriminals have emphasized obtaining enormous volumes of personal information via data breaches in recent years. These hacks expose private information, such as user names, medical information, and login passwords. There are several worries about privacy and data security since the stolen data is often sold on the dark web or used for identity fraud.

Attacks using ransomware, in which thieves encrypt victims' data and demand ransom payments to decrypt it, have become a serious and profitable danger. Furthermore, disruptive cyberattacks that target vital infrastructure, including power grids and healthcare systems, have shown the potential for cybercrime to create havoc and injury in the real world.

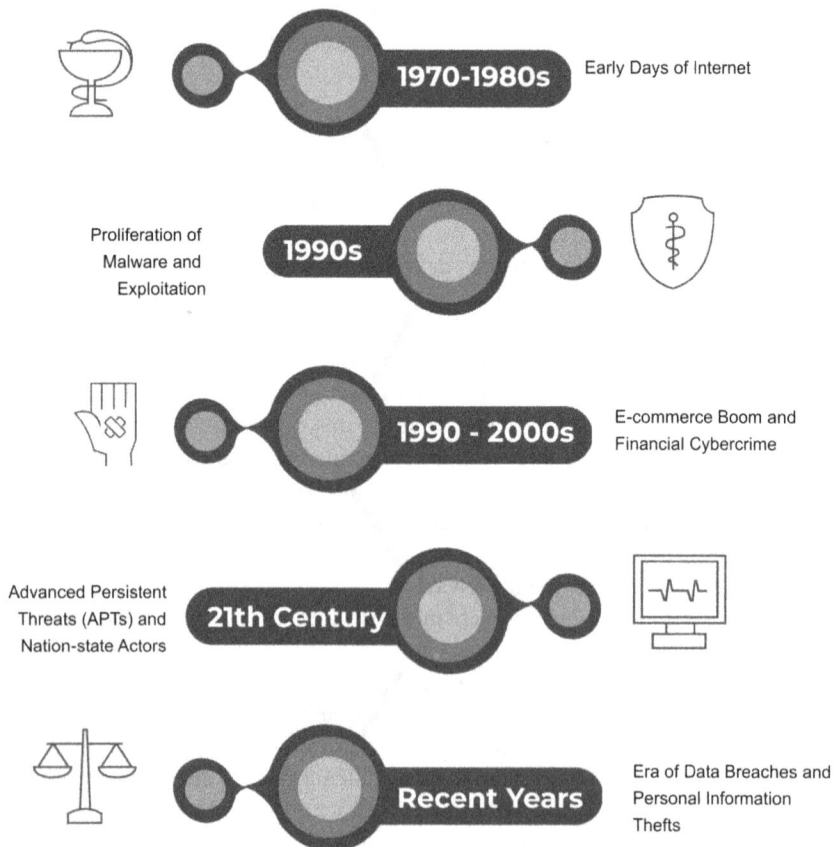

1970-1980s Early Days of Internet

Proliferation of Malware and Exploitation **1990s**

1990 - 2000s E-commerce Boom and Financial Cybercrime

Advanced Persistent Threats (APTs) and Nation-state Actors **21th Century**

Recent Years Era of Data Breaches and Personal Information Thefts

FIGURE 3.1   Evolution of cybercrime

3.1.2 TYPES OF CYBERCRIME

Cybercrime involves various illegal behaviors carried out in the digital sphere. These actions target people and organizations, take advantage of technical flaws, and often have far-reaching effects. Table 3.1 outlines various cybersecurity threats, their concise descriptions, and their potential consequences.

3.1.3 ECONOMIC AND SOCIAL IMPACT OF CYBERCRIME

Beyond the digital sphere, cybercrime is prevalent and has a significant negative economic and social impact on people, companies, and society. The financial toll that cybercrime exacts on organizations and people is one of the most immediate and noticeable effects. Financial theft, ransom payments, and the cost of repairing systems and data are some direct expenses associated with cyberattacks. Long-term damage from indirect expenses, such as loss of consumer trust, reputational harm, and legal obligations, may worsen. Ransomware attacks and high-profile data breaches have caused stock value declines, shareholder lawsuits, and serious business financial losses.

The deterioration of digital trust is a major issue as people depend increasingly on digital platforms for communication, financial transactions, and information exchange. Data breaches and online frauds are examples of cybercrime occurrences that undermine people's trust in the security of their personal information. This deterioration may cause people to be reluctant to use online services, which would slow the expansion of e-commerce and digital innovation. Regulations to strengthen data protection safeguards may alter due to increased privacy concerns.

The effects of cybercrime extend beyond the commercial world and affect both vital infrastructure and public services. Attacks on transportation networks, healthcare systems, and power grids may jeopardize public safety. The possible repercussions include anything from power outages to hacked medical equipment, highlighting the urgent need for effective cybersecurity protections for critical services.

Although often ignored, cybercrime has a considerable psychological cost. Anxiety, despair, and emotional discomfort may occur in people who are the targets of identity theft, cyberbullying, or other forms of online harassment. Cybercrime may breed discomfort in the online community and cause people to behave more cautiously and feel less connected to one another.

Cybercrime has a direct influence on geopolitics in the age of nation-state-sponsored cyberattacks. Cyber espionage and cyberwarfare may undermine international trust, exacerbate political tensions, and strain international relations. A cyber crisis in one nation might have worldwide repercussions due to the interconnection of the digital world.

TABLE 3.1

Types of cybersecurity threats, descriptions, and impact

Cybersecurity Threat	Description	Impact and Consequences
Identity Theft and Financial Fraud	Stealing data to impersonate for financial gain.	Financial loss, damaged credit, emotional distress.
Phishing and Social Engineering	Deceiving for sensitive info; exploiting psychology.	Compromised accounts, ID theft, financial loss.
Cyberbullying and Online Harassment	Online intimidation, spreading falsehoods.	Emotional harm, reputation damage, and conflicts.
Data Breaches and Privacy Violations	Unauthorized access to private data.	Compromised info, legal penalties, trust loss.
Cyber Espionage and Nation-State Attacks	State-sponsored intelligence gathering.	National security risks, diplomatic tensions.
Online Extortion and Sextortion	Blackmail for money or explicit content.	Financial loss, emotional distress, exposure.
Intellectual Property Theft	Targeting patents, copyrights, and trade secrets.	Financial loss, legal battles, and innovation impact.

3.1.4 THE NEED FOR ETHICAL AWARENESS

The significance of ethical awareness cannot be stressed in the quickly changing world of cybercrime, where technology progresses at an unparalleled rate. The ethical ramifications of our online behavior have far-reaching effects as society becomes increasingly linked and dependent on digital networks. Although technology development has brought us incredible ingenuity and ease, it has created new opportunities for abuse and damage. The formation of ethical standards often needs to catch up to the rate of technological advancement, providing gaps that hackers might take advantage of. To foresee and handle any problems, one must thoroughly understand the ethical implications of new technologies.

To ensure that development is in line with the larger welfare of society, ethical concerns should go along with the pursuit of technical innovation. Individuals and organizations are prompted by ethical awareness to consider what should be done in order to reduce damage and promote beneficial results, in addition to what can be accomplished technologically. Finding the correct balance between innovation and ethics is essential for sustainable technological growth.

Technology developers, engineers, and designers have a big ethical job. They can influence security features, privacy preferences, and user interfaces. By adhering to ethical design principles, these experts can produce goods that put user safety, data integrity, and informed consent first. From a technology's initial beginning, ethical awareness urges the inclusion of protections against possible abuse. Individuals are also subject to ethical awareness. Giving people the tools to be digitally literate helps them make better choices, secure their online identities, and contribute to a safer online environment. A culture of ethical online conduct is fostered through understanding the ethical ramifications of disclosing personal information, interacting with online material, and utilizing digital technologies.

Lawmakers and policymakers heavily influence the legal and regulatory frameworks that regulate cyberspace. Ethical considerations must be included while creating these regulations to balance security, privacy, and individual rights. Ethically conscious policies discourage cybercrime, safeguard people's digital rights, and encourage responsible technology usage.

Building ethical consciousness is based on education. People are better prepared to navigate the digital world ethically when ethical and cybersecurity issues are included in school curricula at all levels. Society may jointly contribute to a better, safe, and ethical internet environment by cultivating an early ethical awareness culture.

3.2 ETHICAL ISSUES IN HACKING AND UNAUTHORIZED ACCESS

Hacking, previously linked to technological research and knowledge development, has become a complicated ethical problem with wide-ranging repercussions. Unauthorized access to computer systems and networks presents several moral issues that go beyond the boundaries of technology. This section explores the complex ethical environment around hacking, emphasizing the opposing viewpoints and moral questions.

3.2.1 DEFINING HACKING AND UNAUTHORIZED ACCESS

Although they have different meanings and consequences, the phrases "hacking" and "unauthorized access" are sometimes used synonymously in the context of computer and internet crime. Hacking is using flaws in software, networks, or computer systems to obtain access without authorization, change data, or carry out tasks that the system's creators did not intend. This might include circumventing authentication procedures, hacking security safeguards, and altering code for the desired result.

On the other hand, the act of accessing or utilizing a computer system, network, or online account without the necessary permissions or permission is known as unauthorized access. It includes any effort to cross the limits owners or system administrators placed online. Unauthorized access may entail utilizing stolen credentials, weak passwords, or exploiting software flaws to get access.

While hacking and illegal access may have positive effects, including revealing security holes that can be fixed, they can also potentially do serious damage. Hacking and illegal access are common tactics bad actors use to steal sensitive data, commit fraud, interrupt services, or jeopardize user privacy. The invasion of digital environments and the possible repercussions of these acts create ethical and legal questions.

The motivations underlying the acts are what separate ethical hacking from unethical hacking. Ethical hackers, commonly called "white hat," conduct hacking operations to discover weaknesses and enhance security measures [1]. Together with system administrators, they strengthen defenses against possible threats. Contrarily, unethical hackers, sometimes known as "black hat" hackers, engage in hacking for their benefit, with malevolent purposes, or to do damage [2].

3.2.2   ETHICAL HACKING AND ITS ROLE IN SECURITY

The purposeful and approved effort to penetrate a computer system's security defenses to find vulnerabilities and weak points is ethical hacking, sometimes called "white hat" hacking. In contrast to harmful hacking, ethical hacking is done with express consent to strengthen a company's cybersecurity posture. The integrity, confidentiality, and availability of digital systems and data depend on ethical hacking. Ethical hacking involves skilled individuals who use their knowledge of hacking techniques for positive purposes, safeguarding digital assets from malicious attacks. By identifying vulnerabilities and weaknesses in systems and networks, ethical hackers help organizations address these issues proactively. This includes conducting penetration testing to simulate real-world attacks and assess the effectiveness of security measures. The insights gained from ethical hacking enhance overall security by allowing organizations to refine their strategies, patch vulnerabilities, and strengthen access controls. Ethical hackers educate security personnel within organizations and can be engaged as third-party assessors to provide an unbiased evaluation of security systems. This proactive approach builds user trust and demonstrates a commitment to data protection and business reputation in an era of increasing cyber threats. Figure 3.2 explores the importance of ethical hacking and its multifaceted function in enhancing security.

3.2.3   UNETHICAL HACKING AND ITS CONSEQUENCES

While unethical hacking involves transgressing moral and legal bounds for monetary gain, malevolent purpose, or disruption, ethical hacking helps to find weaknesses and strengthen cybersecurity. Hackers who violate the law are known as "black hat" hackers because they commit cybercrimes that jeopardize the availability, confidentiality, and integrity of computer networks, systems, and data. These actions might have significant repercussions on people, organizations, and society.

Malevolent practices that take advantage of weaknesses and cause chaos in digital environments make up unethical hacking. One aspect is data theft and fraud when nefarious hackers target sensitive data, including people's identities, financial information, and priceless trade secrets. This illegally obtained information opens the door for several criminal activities, such as financial exploitation, identity theft, and the execution of fraudulent schemes. Additionally, unethical hackers spread malware, including various harmful programs, such as viruses, worms, Trojan horses, and ransomware. These cunning tools disrupt digital ecosystems by breaking into networks, stealing data, and creating broad disturbances.

Denial of Service (DoS) assaults, in which dishonest hackers flood systems, networks, or websites with excessive traffic, are another aspect [3]. This tactical inundation essentially paralyzes the targeted resources, making them unreachable to authorized users. As services are disrupted and businesses fight to preserve operational continuity, such assaults leave behind financial losses and reputational damage. In addition, unethical hacking extends to cyber espionage, when hackers carry out covert operations to steal confidential or proprietary information from organizations, companies, or even specific people. Using stolen data for anything from obtaining a competitive advantage to trafficking in illegal markets raises questions about digital security and confidentiality.

Unethical hacking has a wide range of negative outcomes that are catastrophic in many different ways. Organizations struggle most with severe financial losses from breaches, ransom payments, and operational disruptions. However, these financial ramifications go beyond short-term effects and can affect long-term reputational damage and legal obligations. At the same time, individual confidences are violated by unethical hacking, exposing personal information and leading to identity theft, stalking, and unauthorized access to personal accounts. Furthermore, malicious hacking extends outside the digital sphere and impacts larger economies. Numerous hacks ripple across sectors, supply chains, and consumer confidence, eventually causing massive disruptions that fuel economic instability. Businesses also need help with the loss of intellectual property, which includes trade secrets and private information. This double blow results in a loss of competitive advantage and innovation stifling. The effects spread to key infrastructure when immoral hacking reveals flaws in crucial systems like power grids and transportation networks, posing serious risks to operational reliability and public safety.

1

Technical Expertise

Protect digital assets
Identify vulnerabilities

Identify Vulnerabilities

Uncover software weaknesses
Prevent breaches

2

3

Penetration Testing

Simulate attacks
Test defenses

Enhance Security

Patch and reconfigure
Improve access control

4

5

Educate Personnel

Share best practices
Train security teams

Third-Party Assessments

Unbiased evaluations
Insight into weaknesses

6

7

Build User Trust

Demonstrate commitment
Enhance reputation

FIGURE 3.2   Ethical hacking contributions

3.2.4 BALANCING SECURITY AND ETHICAL BOUNDARIES

The pursuit of security measures and the maintenance of ethical limits interact intricately in cybersecurity. Ethical concerns must continue to guide decision-making as businesses and people work to protect their digital assets, sensitive data, and vital infrastructure. The conflict between the need for increased security measures and the preservation of personal privacy is one of the main obstacles to striking a balance between security and ethics. The proper balance must be struck to avoid overreaching into users' lives while maintaining strong protection against cyber dangers. Installing security measures affecting users' privacy must adhere to ethical norms like respect for autonomy and informed consent.

Moral quandaries are brought up using surveillance technology like monitoring software and data-gathering methods. The rights of people to privacy and personal freedom may be violated by the indiscriminate use of these technologies, which may nonetheless help uncover possible security breaches and suspicious activity. Clarity must be established on the scope of surveillance, the reasons for data collection, and the procedures for getting user consent. Transparency and accountability are necessary to maintain ethical standards in security procedures. Organizations should ensure users are informed about the security measures they employ, including the categories of data gathered, how it will be used, and the safeguards in place. Additionally, sustaining trust and exhibiting ethical conduct requires accepting accountability for security lapses or unexpected repercussions.

Unintended effects might result from the goal of strong security. For instance, strict security controls may unintentionally interfere with productive user interactions or inhibit teamwork. According to ethical concerns, security measures must be created with an awareness of possible negative effects on user experiences and the availability of channels for redress if such impacts occur. Human-centered design principles are a moral way to balance security and usability. Users' wants and preferences should be considered while designing solutions, along with aspects like usability, accessibility, and cognitive burden reduction. Prioritizing the human aspect makes security solutions more usable and less likely to violate moral principles. Security ethics must always be reviewed and modified since they are not static. As technology advances, new moral dilemmas emerge. Therefore, to comply with changing ethical standards and legal requirements, organizations and individuals must commit to frequent evaluations of security measures.

3.3 CYBERSECURITY AND ETHICAL RESPONSIBILITIES

Cybersecurity has important ethical obligations in addition to merely safeguarding digital assets. The acts of cybersecurity experts have broad repercussions for people, businesses, and society as a whole in a linked world where digital technologies are heavily ingrained into everyday life. This section outlines the fundamental values that cybersecurity professionals must respect and examines the ethical issues that inform cybersecurity techniques.

3.3.1   THE ROLE OF CYBERSECURITY PROFESSIONALS

Beyond the technological issues, their duties also include preserving moral standards that carefully strike a balance between user liberties and protection. This section explores the many responsibilities of cybersecurity experts and the moral principles that guide their behavior. Cybersecurity experts are entrusted with creating, implementing, and administering security solutions to protect systems, networks, and data from cyberattacks. This entails working with businesses to identify weaknesses, develop risk-reduction plans, and keep up with new threats. These professionals have a variety of tasks, from security analysts to ethical hackers, but they are all motivated by the same goal: securing digital ecologies.

A dedication to moral conduct drives every ethical cybersecurity professional's work. Adherence to values, including confidentiality, integrity, availability, responsibility, and openness, is necessary to protect sensitive information containing personal or private data. Maintaining these principles is essential to preserving user and stakeholder confidence and the security of systems and data. However, this position is challenging. The ethical challenge for cybersecurity experts is to strike a balance between user ease and strong security measures. Achieving this balance requires making judgments that follow security requirements while also considering user demands. The field requires constant study and ethical awareness. Since threats and technology are always changing, ethical cybersecurity experts continue their education to keep current. Additionally, they actively engage in ethical debates inside the industry, working with colleagues to resolve challenges and exchange knowledge.

3.3.2   PROTECTING USER PRIVACY AND DATA

User privacy and data security have become a top ethical issue in the digital era since personal data is regularly traded online. To guarantee the security and confidence of its users, cybersecurity experts and companies managing user data must adhere to ethical standards. Getting informed consent is one of the guiding moral principles in preserving user privacy. Organizations should be open and honest about the data they gather, their motivations for doing so, and their intended uses. Users must be allowed to freely provide their permission and be informed of the consequences of sharing their information. Users should be allowed to opt in or out of various data-gathering techniques via precise and detailed consent.

Data reduction, or gathering just the bare minimum of data required to fulfill an objective, is a technique that is a part of ethical data management. Data that is optional should not be collected or stored. This rule limits the quantity of sensitive data kept, minimizing the possible effects of data breaches and illegal access. Encryption is a vital tool for protecting user data. According to ethical cybersecurity principles, sensitive data must be encrypted both in transit and at rest. This makes it such that even if unauthorized people access the data, they cannot quickly understand its contents.

Additionally, access controls should be built to limit access to data to just authorized individuals. This avoids internal violations by workers or contractors with too many rights. Organizations are required to undertake routine audits and risk assessments as part of their ethical obligation for data security. This entails assessing the security precautions, spotting flaws, and proactively addressing them. Organizations show their dedication to safeguarding user data by regularly reviewing and enhancing security procedures.

Compliance with privacy rules and regulations, such as the General Data Protection Regulation (GDPR) and the California Consumer Privacy Act (CCPA), is a necessary component of ethical data protection conduct [4]. To avoid violating these laws and ethical standards, cybersecurity experts must ensure that their procedures comply with these rules.

3.3.3 BALANCING SECURITY MEASURES AND USER EXPERIENCE

A crucial problem in cybersecurity is striking a precise balance between strong security precautions and a flawless user experience. While strict security rules are necessary to protect private information and digital assets, they may often annoy consumers and make connecting with digital platforms difficult. Maintaining user happiness and ethical obligations depends on finding the correct balance between security and user experience. The task of designing systems that provide thorough defense against possible attacks while ensuring that these controls do not obstruct the usability of the digital environment falls to cybersecurity specialists. They are essential in creating security solutions that are simple to use and intuitive, reducing end users' friction and annoyance.

Achieving this equilibrium is heavily dependent on ethical issues. Integrating security measures with user experience is guided by many important ethical concepts, including transparency, informed consent, and user empowerment. It promotes a feeling of trust and responsibility to make sure users are aware of the security measures in place and how those measures may affect their interactions. Understanding user habits, requirements, and preferences is necessary for developing security measures from a user-centric perspective. This knowledge may direct the development of security features that fit users' mental models, making security precautions seem more like a natural part of the user experience than an impediment.

Both ineffective systems—ones that lack security but are easy to use—are secure. When creating security interfaces and interactions, cybersecurity experts need to take usability and accessibility into account. The combination of user-friendly design, unambiguous instructions, and accessibility features guarantees that security measures are usable by a wide range of users. For the user experience and security measures to be improved, regular user input gathering and iterative design processes are crucial. Maintaining a dynamic security architecture that adjusts to changing conditions is essential as technology advances and threats emerge. User input is an invaluable source of information on improving security without sacrificing usability.

Case Study: Two-Factor Authentication (2FA)

An excellent illustration of how to strike a balance between security and user experience is two-factor authentication. While 2FA considerably improves security, it also adds a second step for users to complete during login. Platforms often provide a variety of 2FA methods, such as text messages, authenticator applications, or biometric verification, to solve this issue, enabling users to choose the one that best suits their tastes and devices.

3.3.4 Transparency and Accountability in Security Practices

Fostering accountability and openness in security processes has become a crucial ethical need in cybersecurity. These principles serve as cornerstones in this rapidly shifting environment, supporting trust while enabling responsible administration of security measures. The open disclosure of a company's security procedures, controls, and incidents is at the heart of transparency. This practice fulfills several vital needs. It first builds user trust by reassuring users of the seriousness with which their data and privacy are treated. Second, openness extends to all parties involved, including partners in business and regulators, displaying a commitment to responsible data management. Transparent communication during security events or breaches aids in efficient situational management by reducing user and stakeholder misunderstanding. Last but not least, open practices demonstrate a dedication to legal standards while assisting in compliance with industry-specific requirements.

Similar to transparency, accountability means keeping people, groups, and organizations accountable for their cybersecurity-related choices and deeds. This accountability is crucial in several important areas. It assigns accountability for handling security breaches and encourages taking preventative action to avoid such incidents in the future. Additionally, accountability fosters a culture of ongoing security measure improvement, where people seek to strengthen defenses and find weaknesses. Accountability helps businesses move toward a future in which security breaches are not only remedied but also their causes are understood and fixed by promoting moral decision-making and reflection after errors.

Organizations may incorporate accountability and transparency into their security policies through certain actions. Building a foundation of trust begins with transparent and ongoing communication of security policies to users and stakeholders. You can ensure quick incident response and promote increased vigilance by providing channels for reporting security issues. Sharing information about events and vulnerabilities openly and honestly may help us learn from them and stop them from happening again. Accountability is important, and effective action-taking and impartial, in-depth investigations into security breaches are key. Roles and responsibilities in cybersecurity should be clearly defined, and training programs may inform staff on how to respect ethical standards and maintain security. Organizations may improve their cybersecurity posture and create a safe online environment for all stakeholders by adopting openness and accountability.

3.4   ETHICAL CONSIDERATIONS IN DATA BREACHES AND INCIDENT RESPONSE

Data breaches are an unwelcome reality that enterprises and people must deal with in today's linked digital environment. Sensitive data may be revealed during a data breach, with serious repercussions for both people and companies. From the first discovery of a breach to the following incident response initiatives, ethical issues are critical in handling these situations.

A data breach occurs when unauthorized individuals access private data kept in an organization's systems. Numerous factors, including software flaws, insider threats, cyberattacks, and human mistakes, may lead to these breaches. Regardless of the reason, anyone impacted by the breach risks serious consequences, including financial loss, identity theft, and privacy breaches.

3.4.1   THE IMPACT OF DATA BREACHES ON INDIVIDUALS AND ORGANIZATIONS

Data breaches have become a common concern in today's linked digital environment, greatly affecting both people and businesses. These incidents happen when unauthorized people access sensitive or secret information, often exposing personal information, financial data, confidential company information, etc. Data breaches have far-reaching consequences beyond simply the initial breach, impacting many areas of people's lives as well as the operations of companies.

Privacy violations are among the worst and most direct effects of data breaches. Organizations get personal information about people, such as names, addresses, social security numbers, and financial information. This sensitive information may be exposed in a breach, resulting in identity theft, fraud, and other cybercrimes. Individuals who are harmed by the loss of privacy may experience mental hardship and financial strain as they deal with the fallout from the violation of their personal information.

Financial damages from data breaches often affect both people and corporations. Financial information that cybercriminals have disclosed may be used to carry out illicit activities, empty bank accounts, or make false purchases. Because of the expenditures involved in detecting and managing the breach and the legal and regulatory consequences, a data breach may result in firms suffering immediate financial losses. Additionally, a decline in consumer loyalty and lower income might result from a lack of customer trust.

A data breach may damage a company's image, harming customer trust and brand impression for years. Rapid media and internet platform dissemination of breach-related information may harm public opinion and undermine consumer trust. Even when organizations take prompt and decisive action to address a breach, rebuilding their image may still be difficult. Data breaches often trigger legal and regulatory repercussions. Data protection regulations in many countries mandate that businesses secure sensitive information and notify the appropriate authorities and anyone impacted by data breaches. Significant penalties and legal action may be imposed for breaking these rules. Affected parties may file legal actions against organizations to demand damages for the harm caused by the breach.

The effects of a data breach might interfere with an organization's regular business activities. Remediation actions, such as looking into the breach, finding vulnerabilities, and putting security measures in place, may take time and resources away from key company operations. The pressure on an organization's IT and cybersecurity personnel to resolve the attack quickly while maintaining data integrity might reduce productivity. After a data breach, confidence may be difficult to regain. People could be reluctant to provide the organization with their personal information if they think the breach will happen again. Organizations must make a concerted effort to convey their priority to data security and privacy, showing their commitment to avoiding future breaches and safeguarding the interests of their stakeholders.

3.4.2 Ethical Handling of Data Breach Incidents

After a data breach, it is critical to handle the situation ethically to minimize damage, ensure transparency, and retain impacted parties' and stakeholders' confidence. Prompt and open communication, aiding individuals in need, and teamwork to guarantee a thorough incident response are all components of ethical responses.

Organizations must first prioritize open communication, swiftly disclosing the scope of a breach and the actions being taken to address it. This openness fosters trust and gives those impacted the opportunity to choose their security in an educated manner. Offering assistance, such as tips on avoiding identity theft and credit monitoring services, displays an ethical commitment to assisting people in navigating the fallout from a breach with the least damage.

Cooperative incident response and evidence retention are part of ethical handling. Cross-functional teams made up of IT professionals, legal counsel, public relations experts, and representatives from the appropriate regulatory agencies should be put together by organizations. By working together, we can comprehensively resolve the breach and preserve the data in a forensically sound way, which is crucial for any legal proceedings and regulatory compliance.

Moral behavior goes beyond the current situation. Organizations should strongly emphasize a culture that values learning and perform post-incident studies to find weaknesses and vulnerabilities. Organizations may proactively strengthen their security measures by implementing enhancements and empowering ethical decision-making at all levels, indicating a commitment to long-term ethical responsibility.

3.4.3 Communication and Transparency in Incident Response

Open and honest communication with stakeholders is essential for sustaining trust, minimizing reputational harm, and ensuring legal compliance. This includes impacted people, consumers, workers, and regulatory agencies. The negative effects of the breach are lessened by prompt and transparent notification, which also shows an organization's dedication to moral and responsible conduct.

Organizations should have a clear communication plan in place in case of a data breach. This plan describes the communication routes, the information that will be communicated with stakeholders, and how the company will alert them. An essential component of incident response is the timely communication of impacted people. People have a right to be informed when their data has been hacked, so they may take the appropriate precautions, such as changing their passwords often, keeping an eye on their bank accounts, and being on the lookout for phishing scams. It is crucial to be transparent when revealing the specifics of the incident. Organizations should be open and honest when discussing the issue, including how it happened, the scope of the breach, and the solutions being used. While there may be worries about possible reputational harm, hiding or downplaying facts may worsen the effects and further undermine confidence. Transparency indicates responsibility and a desire to make things right.

It is crucial to be transparent when revealing the specifics of the incident. Organizations should be open and honest when discussing the issue, including how it happened, the scope of the breach, and the solutions being used. While there may be worries about possible reputational harm, hiding or downplaying facts may worsen the effects and further undermine confidence. Transparency indicates responsibility and a desire to make things right. Media coverage of data breaches is common, which may broaden the impact of the incident. Organizations should collaborate closely with their public relations departments to properly handle media inquiries. Providing dependable information to the media aids with narrative management and helps stop the spread of false information. An open and cooperative approach to media relations may enhance a more favorable view of the organization's reaction.

Additionally, legal and regulatory considerations must be considered while communicating about a data breach occurrence. Many countries require the timely notification of data breaches to the appropriate authorities and the impacted parties. Organizations must ensure that their communication activities comply with these requirements to prevent further fines or legal issues.

Organizations should keep in touch with stakeholders beyond the first disclosure of the breach as the incident response develops. Rebuilding confidence may be aided by regular updates on the status of the inquiry, mitigating circumstances, and proactive initiatives. It may take time and persistent work, but demonstrating continued dedication to security and privacy may ultimately restore confidence. In conclusion, openness and communication are essential elements of moral incident reaction. Effective communication lessens the negative effects of a data breach and shows stakeholders how a business can conduct itself ethically.

3.4.4 LEARNING FROM INCIDENTS FOR FUTURE PREVENTION

Cybersecurity events and data breaches serve as sobering reminders of businesses' changing challenges in the digital era. Even while the immediate aftermath of an event may be difficult, it also gives companies a chance to reflect on their errors, fortify their defenses, and improve their future planning. Learning from events requires a comprehensive strategy that considers technology, procedures, and organizational culture in addition to merely resolving the technical elements. It is critical to carry out a comprehensive post-incident study after a cybersecurity or data breach. This study entails breaking down the event to identify its underlying causes, points of entry, and exploitable vulnerabilities. Organizations may prioritize their efforts to patch vulnerabilities, enhance security measures, and reduce the likelihood of future events by recognizing these characteristics.

An event often exposes a breach in an organization's security posture. This may include issues with staff training, network monitoring, intrusion detection systems, and access control procedures. Organizations may prioritize expenditures in the areas that need the greatest improvement by recognizing these gaps. Penetration testing and regular security assessments may assist in proactively identifying vulnerabilities and shortcomings. The improvement of incident response strategies should follow lessons learned from occurrences. Organizations should assess the effectiveness of their response strategies during the disaster and pinpoint development opportunities. This might include revising communication procedures, defining roles and duties, and updating contact lists. In the case of future events, a successful incident response strategy may reduce damage and accelerate the recovery process.

A common cause of cybersecurity issues is human mistakes. The likelihood of breaches may be significantly decreased by teaching staff members about security best practices, spotting phishing efforts, and emphasizing the value of data protection. Establishing a corporate culture prioritizing security motivates staff to be watchful and proactive in protecting sensitive data. Organizations must stay on top of the ever-changing cyber threat environment. User activity, system logs, and network traffic may all be continuously monitored to spot suspect behavior early on. An organization may stay resilient to new threats by upgrading security policies and procedures based on fresh threat data and lessons learned from prior occurrences.

The exchange of knowledge and cooperation within the cybersecurity sector is essential. In order to keep up to date on the most recent dangers and mitigation techniques, organizations should take part in industry groups, forums, and information-sharing platforms. Other organizations may learn from and strengthen their defenses by exchanging knowledge and experiences from prior instances. Ethical issues should be part of incident learning. Organizations should consider the incident's ethical ramifications, including how it may have harmed stakeholders and those directly impacted. Making ethical decisions includes fixing technological flaws and ensuring people's rights and privacy are upheld throughout the incident response procedure.

3.5 CYBERBULLYING AND ONLINE HARASSMENT

Although the advent of the digital era offers many advantages, it has also given birth to fresh problems like cyberbullying and online harassment. These harmful internet habits may significantly have a detrimental effect on people's mental health, social interactions, and general quality of life. To address these problems, a multifaceted strategy that encompasses people, communities, internet platforms, and law enforcement is necessary. A strong sense of ethical responsibility must drive this strategy.

Cyberbullying is the purposeful harassment, intimidation, or humiliation of a person using digital communication methods, including email, instant messaging, and social networking. It often entails repetitive, damaging encounters that are unpleasant to the emotions. A wider variety of harmful online actions, such as hate speech, doxxing (disclosing private information), and disseminating false material to hurt, are all included in the definition of online harassment.

Cyberbullying and online abuse may have a significant psychological impact. Increased tension, worry, sadness, and even suicide thoughts are common among victims. Internet anonymity may give attackers more confidence, escalating the damage done. Additionally, since online connections are ongoing, victims may find it difficult to flee the suffering, which may result in feelings of helplessness and isolation.

3.5.1 Ethical Responsibilities of Online Platforms and Users

Online communities are used as a means of communication, idea sharing, and debate in the digital era. However, increased connectedness also carries the risk of harmful practices like online harassment and cyberbullying. In order to establish and preserve a secure and courteous online environment, both online platforms and their users have important ethical obligations. As the guardians of their digital communities, online platforms have a huge ethical duty to ensure user contentment. It is essential to carry out several important duties in this position (Table 3.2). Start by setting clear, visible community standards that expressly exclude harmful actions like cyberbullying, harassment, and hate speech. These rules must be conveyed throughout onboarding and readily available to users. To promote user trust and effectively handle inappropriate behavior, it is crucial to operate with consistent and equal enforcement of these guidelines. It is crucial to provide effective reporting systems that give users an easy way to report incidents of cyberbullying or harassment. Platforms should prioritize quick evaluations and respond to these complaints with the required steps. Another crucial responsibility is to ensure that content moderation procedures are transparent. To do this, platforms must provide users access to information about how content removal or account suspension decisions are made and channels for challenging these decisions when appropriate. In addition to combating cyberbullying, platforms must safeguard users' privacy by not disclosing private communications or personal information without permission. Finally, platforms must provide users who are victims of cyberbullying or harassment with services and help. This involves offering data on counseling services, mental health resources, and, if necessary, legal aid.

The ethical behavior and digital citizenship principles adopted form the basis of a polite and beneficial online community. Users should think about several crucial aspects, including starting conversations with respect and empathy similar to real-world interactions, quickly reporting hate speech or cyberbullying using the platform's reporting tools, and standing up to harassment by assisting victims and reporting offensive content. Furthermore, ethical material sharing is a cornerstone that requires users to consider possible effects prior to dissemination. Maintaining others' privacy is also important, requiring avoiding disclosing personal information and the damaging practice of doxxing. Users serve as role models by actively encouraging respectful conduct online and reporting incidents of harassment to maintain a safe online environment.

A comprehensive strategy to combat and prevent cyberbullying is shown in Figure 3.3 to create a more secure and civil online community. The strategies include implementing accessible reporting mechanisms with prompt responses, utilizing cutting-edge content moderation tools that respect freedom of expression, introducing empathy-focused programs encouraging supportive behavior and discouraging harassment, collaborating with law enforcement agencies for serious cases to ensure victim safety and holiness, and many other measures.

TABLE 3.2
Ethical obligations for online platforms

Ethical Obligations	Description	Importance
Community Guidelines	Clear rules against harmful behavior are accessible to users during onboarding.	High
Consistent Enforcement	Fairly enforcing rules to maintain user trust.	High
Reporting Mechanisms	Easy reporting and timely action on cyberbullying.	High
Transparency	Openness on content removal and appeal processes.	Medium
Privacy Considerations	Protecting private info, even when dealing with harassment.	Medium
Support Resources	Providing help for harassment victims.	High

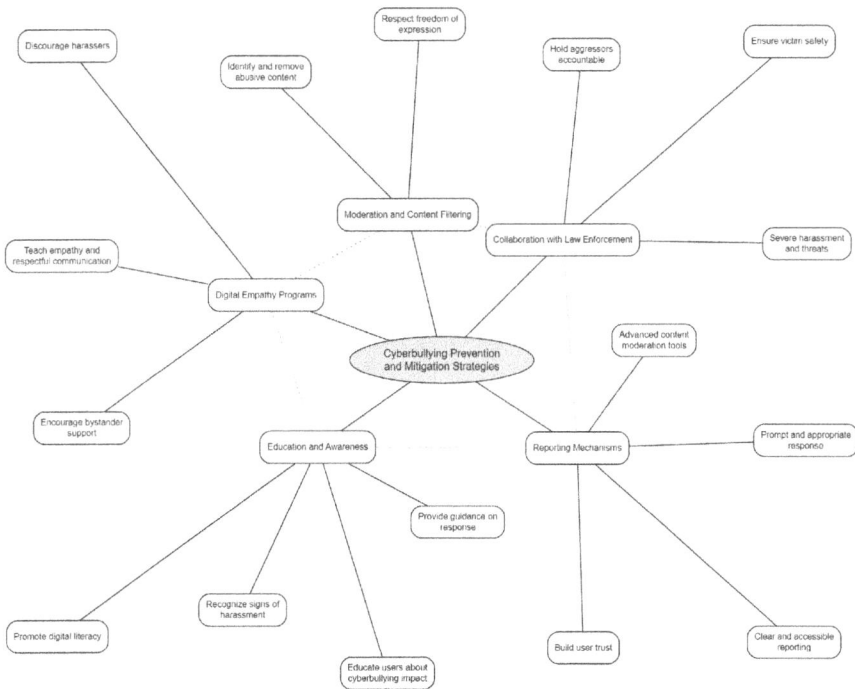

FIGURE 3.3   Strategies for preventing and mitigating cyberbullying: creating a safer online environment

3.5.2   Positive Online Behavior and Digital Citizenship

The prevalence of online harassment and cyberbullying in the modern digital environment has highlighted the critical need for encouraging good online conduct and developing a culture of responsible digital citizenship. It is essential to place a strong emphasis on moral behavior and courteous online interaction in order to address this problem properly. Digital citizenship encompasses a set of attitudes and practices that allow people to navigate the online world responsibly and empathetically. It goes beyond only being technically proficient. Through thorough education, we can enable people to acquire the knowledge and abilities required to interact wisely and respectfully in digital settings. Teaching critical thinking to analyze information sources, building empathy to comprehend many points of view, and promoting ethical conduct while protecting individual privacy are all part of this.

A generation of responsible digital citizens is largely shaped through educational activities. The distribution of dangerous material may be stopped by giving children the skills to distinguish between reliable and false information via integrating digital literacy into the school curriculum. Parents are equally important in supervising their children's online interactions and promoting a positive online profile. Online platforms must actively promote a secure and inclusive digital environment. Instances of harassment and cyberbullying may be quickly addressed by effective moderation procedures, which can also result in penalties for offenders. Platforms may encourage users to participate constructively by rewarding and recognizing good contributions. Furthermore, encouraging bystander intervention may encourage others to take action when they see online abuse, resulting in a coordinated effort to protect the integrity of the digital environment.

A variety of tactics are outlined in Table 3.3 to promote ethical behavior in the online world. The efforts, which include various important acts, are divided into two main categories: "Educational Initiatives" and "Creating Positive Online Spaces." Schools should implement digital literacy courses to give pupils the skills for responsible online involvement. They can also encourage open communication between parents and children about online conduct and encourage platform responsibility via reminders and instructional resources. The table also emphasizes the need for strong moderation mechanisms for quickly resolving online abuse, including gamification components to recognize good efforts and promote bystander action to combat harassment.

TABLE 3.3
Initiatives for fostering positive online behavior

Initiative Category	Initiative	Description
Educational Initiatives	Digital Literacy Programs	Schools integrate curricula teaching responsible online navigation, critical information evaluation, and respectful interaction.
	Parental Guidance	Parents discuss online behavior with children, emphasizing empathy, privacy, and responsible sharing.
	Online Platform Responsibility	Social media and online platforms include prompts, reminders, and educational material for positive online interactions.

(Continued)

TABLE 3.3
Continued

Initiative Category	Initiative	Description
Creating Positive Online Spaces	Moderation and Reporting	Platforms implement robust moderation systems for addressing harassment and abuse and user-friendly reporting mechanisms.
	Rewarding Positive Behavior	Gamification elements reward users for adhering to guidelines and making positive contributions.
	Promoting Allies	Encourage bystanders to intervene in harassment cases; platforms offer tools for supporting victims and reporting harm.

3.6 ETHICAL APPROACHES TO DIGITAL FORENSICS AND CYBER INVESTIGATIONS

Cyber investigations and digital forensics are essential to stopping cybercrimes, finding digital evidence, and maintaining the accuracy of digital data. For these procedures to be fair, accurate, and respectful of people's rights throughout the investigation, there must be a strong commitment to ethical values. The gathering, preservation, analysis, and presentation of electronic evidence in a legally acceptable way are all part of digital forensics. This area is crucial for learning the truth about cybercrimes and ensuring justice is done. Cyber investigations have a wider focus, including identifying cybercriminals and learning about their motivations and techniques.

3.6.1 ENSURING INTEGRITY AND OBJECTIVITY IN DIGITAL FORENSICS

To achieve accurate and trustworthy investigation results, it is crucial to maintain the impartiality and integrity of digital forensics. In this sense, integrity protects digital evidence from illegal adjustments while it is being collected, stored, and analyzed. Contrarily, objectivity necessitates maintaining a neutral, impartial perspective throughout the study, free from any personal prejudices that can skew the results.

A clear chain of custody and rigorous documentation are necessary to accomplish these aims. Every inquiry stage is tracked in detail, fostering accountability and openness. The inquiry's credibility is further enhanced by a carefully established chain of custody that creates a trail of evidence handling from collection to analysis. Furthermore, it is crucial to employ approved instruments and methods. Utilizing reputable and recognized digital forensic tools and techniques provides uniformity and dependability in gathering and analyzing evidence while reducing the possibility of mistakes.

Upholding honesty and objectivity depends heavily on collaboration and peer review. Another degree of validation for the study is added by allowing another expert to assess the procedures and results. Integral to this procedure are ethical issues, notably the harmony between private rights and investigative demands. Making moral choices that uphold people's rights while carrying out investigative duties is crucial for enhancing digital forensics' general legitimacy and dependability.

3.6.2 Balancing Privacy Rights and Investigative Needs

While digital investigations are crucial for revealing evidence and securing justice, they must be carried out with a vigilant awareness of privacy violations. Defining the scope of the inquiry, limiting the amount of data collected to what is necessary, and getting permission or warrants when required are all factors in upholding privacy rights. Even while seeking justice, ethical investigators understand that privacy is a fundamental human right that must be protected.

On the other hand, it is necessary to pay attention to law enforcement requirements and investigate. Situations involving acute dangers to national security or public safety highlight the need for quick data access, obfuscating the distinction between societal protection and privacy preservation. In criminal investigations, digital evidence often plays a crucial role, needing rigorous processes that respect privacy and legal requirements. The key to creating accountability and allaying worries about possible abuse of investigative power is transparency in technique and data collecting.

Principles of proportionality, need, and accountability serve as the moral compass for striking a difficult balance between private rights and the demands of investigative work. In order to maintain the proportionality of invasive measures, the methodology and data that have been gathered must be in line with the seriousness of the current situation. To prevent excessive encroachment into private domains, it is crucial to justify the need for data collection. Accountability is key, so investigators record their processes and judgment calls and stand ready to back up their claims via proper channels.

3.6.3 Ethical Challenges in Handling Digital Evidence

The ethical complexities of handling digital evidence need careful consideration to preserve the objectivity of investigations and safeguard people's rights. Digital evidence must be kept original, and stringent chain of custody rules must be followed from collection to presentation in court. Intentional or unintentional alterations or contamination of digital evidence compromise investigations and diminish public confidence in the legal system.

Investigative needs must be balanced with people's right to privacy, which demands careful ethical navigating. Even though digital evidence may provide important information, investigators must respect private rights and adhere to the law while gathering and managing evidence. Accessing sensitive information without authorization risks violating people's privacy and bringing moral and legal repercussions. Ethical digital forensic techniques are built on transparency and open communication principles. Repeatability is enabled, and the validity of results is ensured through thorough documentation of approaches, tools, and procedures. To disprove accusations of bias in court proceedings, full transparency of the gathering and analysis procedures is essential.

Contextually astute interpretation of digital data is necessary to prevent forming false or deceptive inferences. In order to avoid confirmation bias and excessive reliance on particular types of evidence, ethical practitioners should strive toward impartial presentations that take uncertainty into account. Experts in digital forensics routinely testify in court, highlighting their moral obligation to the judge. It is crucial to convey results objectively and honestly. Misrepresentation or unsubstantiated assertions may damage the credibility of both the expert and the evidence.

3.6.4   Admissible Evidence and Legal Considerations

The integrity of the evidence is crucial in determining its admissibility since it guarantees that it complies with legal requirements and upholds individual rights. The requirements for admission often center on relevance, genuineness, and adherence to the hearsay rule due to the diversity of jurisdictions. Digital evidence must be able to prove its provenance, unaltered nature, and context in order to be accepted as legitimate.

Digital evidence, however, has its own set of difficulties. More attention is required throughout data collection, preservation, and analysis to avoid data manipulation or contamination due to the malleability and complexity of digital environments. A major challenge is authentication, which requires identifying the evidence's origin, user identity, and creation conditions in order to vouch for it as authentic and unmodified. These difficulties are exacerbated by encrypted data, which also raises moral and legal concerns about privacy rights and decryption key access.

Ethical issues highlight the precarious balance between justice and privacy in this environment. As investigators negotiate the moral conundrums posed by handling personal data, upholding individual rights becomes crucial. The investigation procedure should be transparent from data collection through analysis to increase confidence and strengthen the validity of the evidence. As ethical requirements, the impartiality and objectivity of digital forensic professionals also arise to prevent prejudices that can jeopardize the inquiry and the admissibility of evidence.

3.6.5   The Role of Transparency and Accountability

In digital forensics and cyber investigations, transparency and accountability are fundamental principles that act as moral protections to maintain the objectivity of the inquiry. Transparency is the open and honest sharing of methodology and conclusions throughout the inquiry, strengthening credibility and allowing for outside evaluation of approaches used. By strengthening the public's perception of the validity of the findings, this method fosters trust and confidence in the investigation process. Accountability mandates that investigators remain unbiased and adhere to ethical standards while acknowledging and addressing any possible biases.

These ideas only come with difficulties. Transparency initiatives may be made more difficult by the complexity of technology, concerns about confidentiality, and the need to strike a balance between openness and the preservation of sensitive information. Finding the ideal balance often requires consulting impartial reviewers or outside specialists to evaluate procedures and results. Education and training are essential to ensure that investigators have a thorough knowledge of these principles and promote a culture that promotes accountability and openness.

3.6.6   Continuous Learning and Adaptation

Due to the quick development of technology, adjustments in cybercriminals' strategies, and revisions to regulatory frameworks, digital forensics and cyber investigations are always changing. The highest ethical standards must be maintained to maintain effectiveness, and ethical practitioners in this sector know the need for constant learning and adaptation.

The extraction, analysis, and preservation of digital evidence largely rely on technology in digital forensics. New gadgets, file formats, and communication channels emerge as technology develops. To grasp these new technologies and their possible consequences for investigations, professionals in the sector must actively participate in continual learning. Regular training sessions, seminars, and conferences provide opportunities to learn about recent technologies and procedures.

Countries may have different laws and rules regarding using digital evidence in cyber investigations. Ethical digital forensics practitioners must remain current on legislative changes and court rulings that influence their work. Understanding privacy regulations, evidentiary standards, and court rulings concerning the admissibility of digital evidence are all part of this. An accurate knowledge of legal issues might lead to improper evidence management or violations of people's rights.

New research fields have emerged due to the growth of digital spaces, including those related to social media platforms, IoT (Internet of Things) devices, and encrypted communications. These new fields raise particular ethical questions. To ensure they follow ethical guidelines, professionals must proactively address concerns about privacy, consent, and the use of new technology. Working with legal professionals and stakeholders may be necessary to address these issues.

Continuous learning in digital forensics involves individual efforts, group work, and information exchange within the industry. Professionals may share experiences, ideas, and best practices via forums, online groups, and trade organizations. Practitioners jointly contribute to the evolution of ethical norms in the profession by exchanging ethical issues and lessons learned. The ethical standards that govern digital forensics may need to change as the profession develops. Digital forensics' ethical issues may be intricate and multidimensional, and keeping up with changing standards is crucial. To guarantee that policies remain applicable and useful, professionals should participate in conversations around ethical codes of conduct and share their thoughts.

REFERENCES

1. Hatfield, J.M., Virtuous human hacking: The ethics of social engineering in penetration-testing. *Computers & Security*, 2019. 83: pp. 354–366.
2. Silic, M. and P.B. Lowry, Breaking bad in cyberspace: Understanding why and how black hat hackers manage their nerves to commit their virtual crimes. *Information Systems Frontiers*, 2021. 23: pp. 329–341.
3. Arora, A., S.K. Yadav, and K. Sharma, Denial-of-service (dos) attack and botnet: Network analysis, research tactics, and mitigation, in *Research Anthology on Combating Denial-of-Service Attacks*. 2021, IGI Global. pp. 49–73.
4. Alexander, C.B., The general data protection regulation and California Consumer Privacy Act: The economic impact and future of data privacy regulations. *Loyola Consumer Law Review*, 2019. 32: p. 199.

4 Privacy
Ethical Dimensions in the Digital Age

4.1 THE IMPORTANCE OF PRIVACY IN THE DIGITAL ERA

The value of privacy cannot be emphasized in the digital era when technology is seamlessly integrated into every aspect of our lives. Not just a basic human right but also the cornerstone of a free and democratic society, privacy is not only a luxury. Our private information, communications, and actions are more exposed than ever as our lives grow increasingly entwined with digital platforms. Privacy protection has become a crucial ethical and cultural issue due to the significant problems and enormous advantages this digital change has brought.

4.1.1 PRIVACY IN THE INFORMATION AGE

The idea of privacy has undergone a profound transformation due to the unparalleled technical development and interconnectedness of the Information Age. The protection of privacy in this digital age is a crucial issue as well as a difficult task. The digital economy is now driven by the acquisition, analysis, and use of data, which has become a valuable asset. The volume of data collecting, the openness of data methods, and the nature of user permission are just a few of the complex ethical issues this data explosion has brought to light.

Widespread monitoring, made possible by digital gadgets and cutting-edge technology, characterizes the Information Age. Governments and commercial organizations use security cameras, face recognition software, and smartphone monitoring to keep an eye on people. They present issues with civil liberties and personal privacy while boosting security. At the same time, anonymity has become harder to achieve. Through online interactions, people often leave digital traces that may be exploited to identify them. Free speech and self-expression may be discouraged by this loss of anonymity.

A key ethical dilemma is balancing the benefits of technology advancement and privacy protection. Strong legal frameworks, ethical data practices, and an educated public are necessary for achieving this balance. It is crucial to have conversations about the boundaries of surveillance, the need for data gathering, and the ethics of data use. The Information Age forces society to redefine and protect privacy, realizing the necessity for constant adaptation in a time of fast technological advancement.

DOI: 10.1201/9781003584452-4

4.1.2 EVOLUTION OF PRIVACY CONCERNS

The fast growth of technology and the evolving nature of how people interact with information and one another have significantly impacted privacy issues in the digital age (Figure 4.1). Understanding the intricate ethical issues surrounding privacy today depends on this growth. In the past, privacy issues were mostly related to actual places and real papers. People were concerned that snoopers would listen in on their private correspondence, diaries, or talks behind the walls of their houses. Protecting the sanctity of one's physical space and material goods was at the heart of privacy.

However, as information technology, notably the internet, developed, the emphasis switched from defending physical areas to defending digital assets. This change brought up fresh concerns regarding data security and the management of personal data. People suddenly had to deal with the repercussions of revealing private thoughts, photos, and conversations on digital platforms like email, social media, and instant messaging services.

Big data analytics and data mining tools have created fresh privacy worries. Large datasets might now be gathered and analyzed by businesses and groups to provide comprehensive profiles of people. The use of this profiling for targeted advertising aroused concerns about possible discrimination, spying, and the erosion of personal freedom. The ethical issues around data collecting, use, and protection emerged along with privacy concerns. In the digital sphere, problems of permission, transparency, data ownership, and responsibility are now being debated by individuals, organizations, and governments.

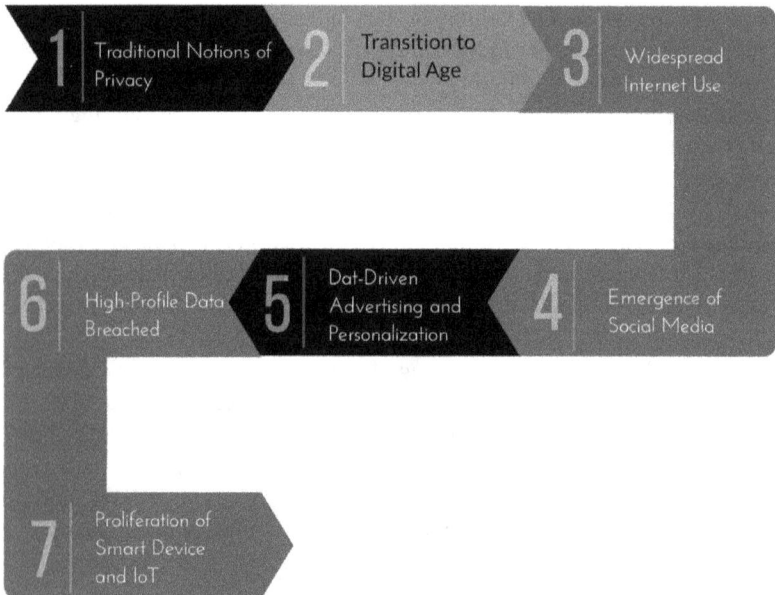

FIGURE 4.1   Evolution of privacy concerns in the digital age: A simplified overview

4.1.3 Privacy as a Fundamental Right

Privacy is a basic right that is founded in the very nature of human liberty and dignity. This right, recognized by international treaties and legal frameworks, emphasizes that people naturally can regulate how their personal information is collected, used, shared, and retained. It encompasses various aspects of life, including private conversations, medical information, and financial data. In essence, privacy protects human dignity by providing defense against the degrading impacts of ongoing monitoring.

People may express themselves, build connections, and explore their opinions and wants without being constantly watched in a culture that cherishes privacy. It serves as a vital check against the abuse of power, preventing governmental bodies, businesses, and other institutions from going too far. People are more willing to engage in online shopping and healthcare activities when they are certain that their personal information will be respected and protected. The underpinning of contemporary economies and fundamental services is this trust. Instead of restricting creativity or freedom of speech, privacy often fosters both by guaranteeing people that their thoughts and creative works will be shielded from undue examination. Therefore, privacy promotes an innovative and expressive culture. Maintaining privacy as a basic right in the digital age is crucial for the moral operation of society, preserving individual liberty, human dignity, and trust.

4.1.4 Privacy and Personal Freedom

Personal freedom and privacy are intimately linked; this relationship is at the core of each person's autonomy and self-determination. This connection is more important than ever in the digital age since how our personal information is gathered, utilized, and processed significantly impacts our freedoms and choices. Personal freedom is fundamentally the capacity to make decisions about one's behaviors, beliefs, and way of life without the influence of others. A crucial aspect of preserving this liberty is privacy. People may decide what information to reveal, with whom to share it, and under what conditions when they have control over it. They may mold their online interactions and identity with this option to reflect their tastes and ideals.

Having privacy helps create a space to think freely and express oneself. In a future where digital footprints are continually tracked, people may hesitate to voice opposing views or participate in candid conversations. Fear of being watched or of social repercussions may cause self-censorship, stifling innovation, and the free flow of ideas. Individuals may investigate other viewpoints, question the established quo, and have productive conversations in a private setting without always being watched and judged for their words and deeds.

Privacy safeguards that protect individual data may lessen prejudice. Algorithms and machine learning systems often make judgments that impact people's lives in a data-driven society, ranging from employment chances to loan approvals. These technologies may reinforce prejudices and preconceptions without privacy protections, restricting the personal freedom of underprivileged populations. By minimizing such prejudice, privacy laws and moral data-handling procedures enable everyone to achieve their goals and benefit from equitable possibilities.

Particularly in delicate industries like healthcare and counseling, privacy is essential. People should not be afraid to seek medical care, mental health assistance, or counsel without worrying that their private information may be disclosed. Without privacy guarantees, individuals could put off getting care when needed, resulting in health problems and losing their right to privacy.

4.2 PRIVACY RIGHTS, DATA PROTECTION, AND CONSENT

The idea of privacy rights, data protection, and informed permission has taken center stage in debates about digital ethics in an era marked by the persistent collecting and use of personal information. The fundamental values and ideas that guide the ethical aspects of data management and privacy in the digital age are explored in this section.

4.2.1 LEGAL FRAMEWORKS FOR PRIVACY PROTECTION

Legal frameworks for privacy protection have arisen as essential instruments in maintaining people's rights and resolving ethical problems in the digital era, as personal data is continually gathered and processed. They take many different shapes, outlining the parameters and responsibilities of privacy protection in anything from international accords like the Universal Declaration of Human Rights to state laws [1]. A major example is the General Data Protection Regulation (GDPR), which was implemented by the European Union [2] and has considerably influenced international data privacy standards. While imposing significant obligations on enterprises, GDPR unifies the regulations for processing personal data across all EU member states, providing people the right to view, correct, and destroy their data. Similar privacy rights are expanded in the United States by the California Consumer Privacy Act (CCPA), which gives citizens control over their personal information [3].

These frameworks must include cross-border data transfer methods that facilitate data flows while guaranteeing uniform privacy protection requirements. Data Protection Authorities (DPAs) supervise compliance, look into breaches, and provide advice as part of their crucial enforcement function [4]. Industry-specific rules provide particular privacy protections, such as HIPAA for healthcare data. While these legal frameworks are essential for protecting private rights, they also come with difficulties, including the requirement for global harmonization and ongoing adaptation to rapidly changing technology environments. An overview of the major legislative frameworks for privacy protection, including their reach, significant clauses, and enforcement mechanisms, is given in Table 4.1.

4.2.2 GDPR AND ITS IMPACT

Key concepts stressing individual data rights, data minimization, openness, and responsibility are the foundation for this comprehensive EU rule. It upholds the rights of people, including their right to access and amend their data and their "right to be forgotten." GDPR requires the designation of Data Protection Officers (DPOs) and impact analyses, holding corporations responsible for data protection [5]. Importantly, the extraterritorial nature of GDPR makes it a global data protection norm as it requires compliance from enterprises processing the personal data of EU individuals anywhere in the globe.

Organizations have had a significant effect from GDPR. Entities must evaluate their data-processing procedures, put strong data protection measures in place, and use privacy by design principles to comply. Organizations are encouraged to prioritize data privacy since compliance violations may result in significant penalties. Although the GDPR has presented difficulties, like a large data inventory and technological adjustments, it has also produced advantages, including increased customer trust and greater data control.

The impact of GDPR on the world cannot be understated. In response to the rising significance of data privacy in the digital era has motivated other governments and areas to tighten their data protection legislation. The GDPR continues to shape the ethical dimensions of privacy, emphasizing the need for transparency, accountability, and respect for individual data rights across borders and industries as individuals gain more control over their data and organizations worldwide prioritize ethical data handling.

TABLE 4.1

Key legal frameworks for privacy protection

Legal Frameworks for Privacy Protection	Scope	Key Provisions	Enforcement
International and National Legislation	– Global and National	– Varies by jurisdiction	– Varies by jurisdiction
General Data Protection Regulation (GDPR)	EU-wide	– Right to access personal data – Right to be forgotten (data erasure) – Data protection impact assessments	European Data Protection Authorities
California Consumer Privacy Act (CCPA)	State of California	– Right to know what personal data is collected – Right to delete personal information – Right to opt out of data sharing	California Attorney General's Office
Cross-Border Data Transfer Mechanisms	International	– Ensures data protection during cross-border transfers	Data Protection Authorities and Legal Mechanisms
Data Protection Authorities (DPAs)	Varies by jurisdiction	– Enforcement of privacy regulations – Investigation of data breaches – Guidance to organizations and individuals	Various national and regional DPAs
Sector-Specific Regulations	Industry-specific	– Healthcare data privacy standards – Financial data privacy requirements	Relevant industry oversight and regulatory bodies

4.2.3   INFORMED CONSENT IN THE DIGITAL CONTEXT

Informed consent—a core ethical ideal in data collection and use—faces particular difficulties. In the digital world, informed consent sometimes takes the shape of a checkbox buried inside long privacy rules or terms of service agreements that consumers may easily ignore or not fully understand. This raises significant ethical issues. The complexity and legibility of permission processes call for more simplicity and clarity. Users should have no trouble understanding how their permission will be used. Second, the granularity of data gathering in the digital age necessitates real options that let consumers give or withhold permission for specific uses at their discretion. Thirdly, ethical consent should consider that data moves dynamically across a complex ecosystem by allowing individuals to examine and modify their choices as their data is used in various circumstances.

Additionally, ethical procedures must give vulnerable groups like minors, who may need help understanding the implications of data sharing, special attention, guarantee the revocability of permission, and handle third-party data sharing. In response, moral platforms and organizations are implementing user-centric strategies for consent. They want to enable consumers to choose their settings, explain data use clearly and succinctly, and give frequent updates on privacy options. With an emphasis on user autonomy in data processing, regions like the European Union have adopted strict criteria for informed consent under the GDPR.

4.2.4   DATA OWNERSHIP AND CONTROL

Data ownership and control are the core ethical questions in privacy and data management in the digital age. These arguments center on the issue of whose personal data should be legally owned and governed. Data is a valuable resource and a reflection of one's identity; as such, it is a difficult topic that people, businesses, and legal systems struggle with. Personal data ownership is increasingly seen as a fundamental human right. The recognition of people as the principal proprietors of their data is encouraged by ethical ideals. Individuals should have the last say on how data is gathered, used, and shared since data is created and disseminated via interactions with digital platforms, devices, and services. This viewpoint supports that data is a property right, giving people freedom and control over their digital identities.

Data ownership, however, is more than just expressing rights; it also entails good care. The significance of managing data in a manner that respects individual rights and interests is recognized by ethical groups. They prioritize informed consent and allow people to give or withdraw their permission for data collection and processing. Furthermore, new technological developments, like blockchain, give people more control over their data, changing who owns it and how it is used. This dynamic interaction of rights, obligations, and technology underscores the ethical need to find a balance in data ownership and management, protecting privacy, and building trust in the digital age.

4.3   ETHICAL CONSIDERATIONS IN DATA
COLLECTION AND USAGE

As our lives grow increasingly digital, data gathering and utilization have proliferated in the digital age. Although data may spark innovation, provide insightful information, and enhance services, it also creates significant ethical issues. How businesses gather, manage, and use data significantly influences people, society, and even democracy.

4.3.1   ETHICS IN DATA COLLECTION PRACTICES

Ethical data-gathering procedures are the cornerstone of competent data management in the digital era. Transparency, permission, and data reduction are the three main tenets of these procedures, which jointly protect people's privacy and preserve moral principles (Figure 4.2). Informed consent and transparency are essential. Organizations must make their data collection practices, goals, and uses transparent and simple to understand. Informed consent refers to the freedom of people to give or withdraw their assent after full information and without pressure. Additionally, they should be free to change their minds at any moment.

Another important idea is data reduction, which emphasizes gathering the necessary information for the task. In addition to raising ethical questions, gathering excessive or unnecessary information increases the danger of abuse and data breaches. Organizations should avoid collecting data "just in case" and constantly evaluate how well it aligns with their goals. Fundamental to user autonomy is respect. Users must have control over the data they provide, including whether to opt in or out of data collection, the opportunity to choose the precise data to supply, and the capacity to pick the duration of data retention. Strong data security requirements are also necessary for ethical data collecting to guard against unwanted access and breaches.

4.3.2   DATA MINIMIZATION AND PURPOSE LIMITATION

Data reduction and purpose restriction play a crucial role in defining ethical data management practices in data collection and utilization. Data minimization promotes gathering only the absolute minimum of information required to achieve a clearly stated goal. This moral rule reduces the likelihood of privacy violations while making it easier for businesses to safeguard and manage their data efficiently. A weather app, for instance, should only ask for and preserve location data necessary for producing reliable predictions, avoiding gathering extraneous data such as contacts or browser history.

The notion of purpose restriction, which emphasizes that data should be used only for the reasons it was first gathered and to which persons agreed, complements the idea of data reduction. In terms of ethics, this implies avoiding using gathered data for unrelated purposes without obtaining permission or a valid reason. For instance, a social media platform should only utilize customer information collected for social interaction and advertising targeting with the express approval of the customer.

Purpose restriction promotes appropriate data practices, user involvement, and privacy protection while fostering trust between companies and users.

It is impossible to exaggerate the ethical importance of data reduction and purpose restriction. These guidelines reduce the dissemination of personal data and guarantee that it is used only for the objectives for which it was collected, acting as crucial protections for individual privacy. Organizations may increase trust, improve data security, and often abide by data protection laws by upholding these values. Adopting data reduction and purpose restriction in an era when data collecting is on the rise shows a commitment to upholding people's rights and promoting ethical data management.

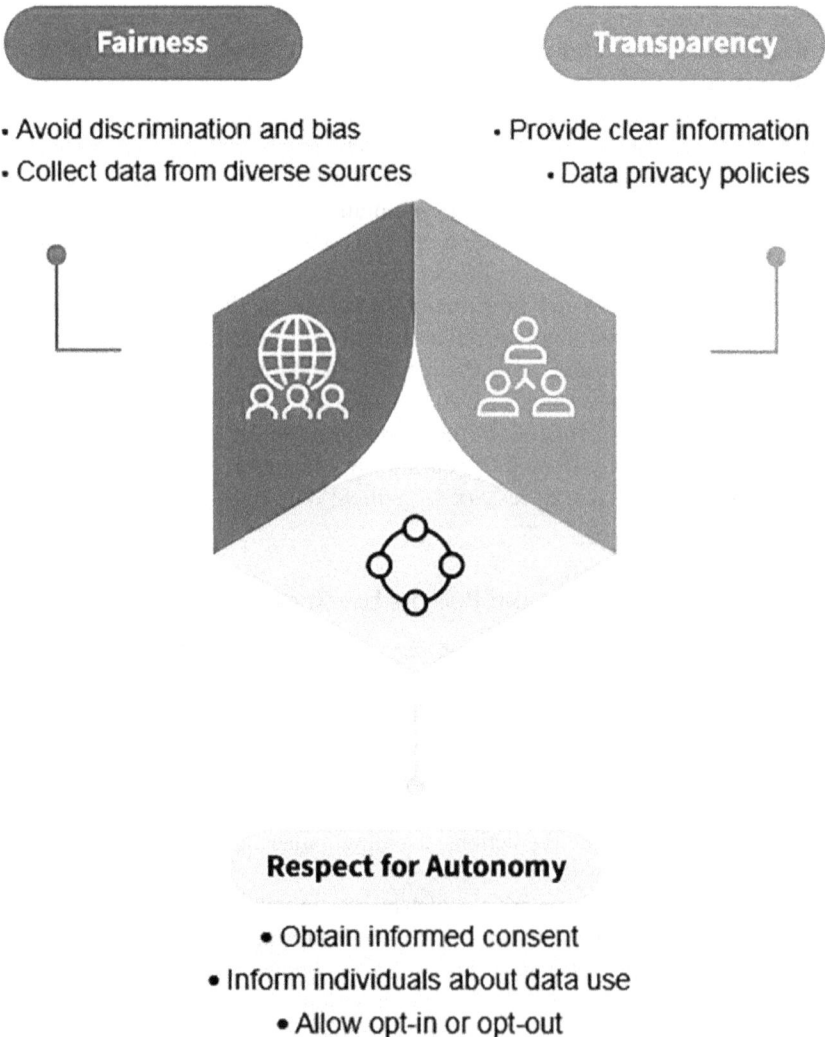

FIGURE 4.2   Ethics in data collection practices

4.3.3 USER PROFILING AND ALGORITHMIC BIAS

User profiling comprises the development of comprehensive personal profiles for each user based on their online activity. This may improve personalization but also raises privacy and discrimination issues. Key ethical issues in this situation are transparency, consent, and non-discrimination, which call for clear communication, informed consent, and the avoidance of discriminatory effects. Conversely, algorithmic bias occurs when artificial intelligence (AI) systems or machine learning algorithms provide unfair or biased findings due to skewed data or underlying assumptions. The main focuses of ethical concerns include reducing prejudice, promoting justice and equality, increasing openness, and creating responsibility. In addition to having negative effects on access to opportunities and systemic prejudices, biased algorithms may exacerbate discrimination, undermine public confidence in technology, and have broad social repercussions.

These moral conundrums emphasize the need for ethical data procedures. Privacy rights are respected by ensuring openness and permission in user profiling, preventing discrimination in profiling, and correcting data bias. Fairness, accountability, and openness are essential to preventing algorithmic bias and promoting faith in digital services and technology.

4.3.4 THE ROLE OF TRANSPARENCY IN ETHICS

A key value for businesses in the digital era is transparency, which is the cornerstone of moral data gathering and utilization. It involves open and transparent communication about data practices, making sure people know the information gathered, how it will be used, and who has access to it. Building confidence between companies and people depends on this openness. It helps protect people's rights and privacy by enabling them to make well-informed decisions about sharing their data.

Clear communication of data practices, gaining informed permission, establishing data usage and sharing rules, explaining data security measures, and giving updates on any practice changes are just a few of the essential components of ethical transparency. Beyond just adhering to the law, it is a dedication to responsibility and respect for personal liberty.

Transparency has ethical relevance by encouraging trust and responsibility, upholding human liberty, assuring compliance with data protection regulations, reducing data abuse, and promoting a socially responsible data culture. Transparency is a basic ethical necessity for companies looking to manage the ethical complexity of data collection and utilization in an era when data is central to our lives. It is a recommended practice and a legal obligation in many places.

4.4 BALANCING PRIVACY WITH SECURITY AND PUBLIC INTEREST

A major ethical and policy challenge today is how to strike a delicate balance between individual privacy, security, and the larger public interest. In a world where threats to people and society may have serious repercussions, it is crucial to maintain individual privacy while also taking security precautions. There are times when the public interest, such as preventing crime or preserving national security, may conflict with a person's right to privacy.

4.4.1 NATIONAL SECURITY VS. INDIVIDUAL PRIVACY

The conflict between the need for national security and the preservation of individual privacy poses a significant ethical and policy dilemma. Governments and security organizations contend that broad surveillance and information-gathering authority is required to protect the country from dangers, including terrorism, cyberattacks, and espionage. However, these initiatives often directly impact the rights and expectations of personal privacy.

Government surveillance operations, including information gathering, communications monitoring, and cutting-edge technology like face recognition, have raised serious concerns about possible overreach. According to critics, the widespread gathering of data on common citizens who are not accused of any crime violates people's right to privacy. The moral conundrum at the center of this discussion is how to strike a balance between national security and individual privacy needs. While it is the responsibility of governments to safeguard their people, the loss of privacy rights may result in power abuses, privacy breaches, and a restraint on dissent and free expression.

Many nations have put in place legislative frameworks and supervision procedures to ensure that surveillance operations are carried out within the boundaries of the law while protecting individual rights to overcome this ethical conundrum. To strike the correct balance between national security needs and individual privacy rights, the government must be transparent in its operations, have clear guidelines for surveillance, and create independent review organizations. As technology develops, ethical concerns must also apply to the appropriate creation and use of surveillance systems, particularly limiting privacy invasions.

4.4.2 SURVEILLANCE AND PRIVACY INTRUSIONS

The moral conundrum posed by monitoring and invasions of privacy has grown in complexity. Data collecting methods used in mass surveillance are often indiscriminate, which raises questions about how innocent people's privacy will be protected. Without their agreement, technologies like face recognition and data mining may identify and profile people, and intrusive data gathering by governments and commercial companies puts the openness and consent important to ethical surveillance in jeopardy. Privacy, openness, and consent are the key ethical issues. Convenience, security, and privacy must be balanced. Surveillance technology, including AI and machine learning, must be developed and used properly to prevent abuse. Whether the government or the private sector monitors, the ethical challenge is to protect basic human rights while addressing legitimate security concerns.

Ultimately, the discussion of monitoring and privacy invasions highlights the need for continual public conversation and democratic decision-making. In order to balance upholding civil liberties with addressing the digital era's real security and social demands, societies must meet the expanding difficulties provided by new monitoring technology.

4.4.3 THE ETHICAL DILEMMAS OF PUBLIC INTEREST

The idea of "public interest" often raises difficult moral conundrums. It may not be easy to define the public interest since many stakeholders, including governments, businesses, advocacy organizations, and people, have different points of view. These divergent points of view may result in serious ethical disagreements about whether activities serve the interests of the public.

The trade-off between security measures and individual privacy rights is a serious ethical conundrum. According to governments, the public interest is served by increased security measures like monitoring and data collecting that shield residents from danger. However, privacy activists claim these procedures violate individual rights, raising worries about losing privacy in the name of security.

The environment of content control and restriction on internet platforms presents another ethical dilemma. Although these steps may be justified as being required to safeguard the public interest by preventing hate speech or misinformation, they may also raise questions about whether they restrict freedom of expression. In these situations, defining what constitutes the public interest might be difficult.

Tracking and monitoring people's activities and health information is essential for maintaining public safety during public health emergencies like pandemics. A careful ethical balance must be struck between privacy rights and health and safety concerns since these measures may infringe on people's rights to privacy and autonomy.

Legal structures, checks and balances, and independent monitoring agencies are crucial to maintain this balance. In order to sustain ethical standards and manage the complex and dynamic difficulties related to public interest and privacy in the digital age, transparency, open debate, explicit norms, and accountability systems play crucial roles.

4.4.4 STRIKING A BALANCE: PRIVACY BY DESIGN

The proactive and comprehensive approach known as "Privacy by Design" has become a key foundation in the continuing ethical discussion around privacy, security, and the public interest. By including privacy issues from the very beginning of systems and processes, privacy by design fundamentally aims to achieve a harmonic balance between these conflicting objectives.

Privacy by Design is a fundamental change in how businesses approach data management and technological development. It goes beyond a compliance checklist. In order to ensure that privacy is not an afterthought or an extra add-on but a basic concept that governs every step of development, it stresses the incorporation of privacy protections into the very DNA of technologies and systems.

Privacy by Design conforms to values like data minimization, informed consent, and openness. It encourages businesses to gather the data required for certain tasks, tell consumers clearly about how their data will be used, and uphold openness in data practices and security measures. A key component of this strategy is privacy impact assessments (PIAs), which enable firms to identify and reduce possible privacy issues before introducing new technologies or procedures.

Although privacy by design presents a potential way to strike a compromise between conflicting interests and respect moral principles, its implementation may take time and effort. Organizations need to change ingrained data habits that put convenience above privacy. Additionally, to guarantee that the concepts of privacy by design are understood and adopted, it is essential to educate stakeholders, including developers, politicians, and users. Table 4.2 illustrates each privacy by design principle, provides practical implementation guidelines, and offers real-world examples showcasing the application of these principles in various contexts.

TABLE 4.2

Illustrating privacy by design principles with examples and implementation guidelines

Privacy by Design Principle	Description	Implementation Guidelines	Examples
Proactive, Not Reactive	Embedding privacy considerations into projects from the outset.	– Conduct privacy impact assessments early in project planning.	– Incorporating privacy safeguards while designing a new mobile app to protect user data.
	Anticipating and preventing privacy issues before they arise.	– Train staff on privacy best practices.	– Identify pot and address potential privacy risks in a healthcare software system natively.
Privacy as the Default Setting	Ensuring that privacy protections are automatically applied by default.	– Design systems to limit data collection to what is strictly necessary for the intended purpose.	– Implement default privacy settings that limit data sharing on a social media platform.
	Users should not be required to take action to protect their privacy.	– Implement strong encryption by default for data in transit and at rest.	– Automatically encrypt emails using an email service without requiring users to enable encryption.
Privacy Embedded into Design	Making privacy an integral part of the system or product design.	– Involve privacy experts in the design process.	– Collaborating with privacy experts to design a secure and privacy-aware IoT device.
	Considering privacy at every stage of development.	– Identify potential privacy risks and constraints in design specifications.	– Ensuring that privacy features are considered from the initial concept phase of a new software application.

(Continued)

TABLE 4.2
Continued

Privacy by Design Principle	Description	Implementation Guidelines	Examples
Full Functionality, Positive-Sum	Striving to deliver all intended functionality without compromising privacy.	– Seek innovative solutions that balance functionality and privacy.	– Developing a healthcare app that provides personalized recommendations without compromising patient privacy.
	Avoiding the misconception that privacy and functionality are conflicting.	– Evaluate the trade-offs between data collection and user benefits.	– Balancing user experience with data protection in a fitness tracking app.
End-to-End Security	Implementing robust security measures to protect data throughout its lifecycle.	– Conduct regular security audits and assessments.	– Implement end-to-end encryption in a messaging app to ensure secure user messages.
	Addressing security and privacy comprehensively and consistently.	– Encrypt sensitive data during storage, transmission, and processing.	– Storing healthcare records securely in a medical database and encrypting data in transit.
Visibility and Transparency	Maintaining openness about privacy practices, policies, and procedures.	– Communicate the purposes and uses of collected data.	– Providing users with a clear and concise privacy policy that explains data usage in a cloud storage service.
	Providing users with clear information about data collection.	– Allow users to access and review their data.	– Allowing users to view their location history in a mapping app.
Respect for User Privacy	Prioritizing user privacy preferences and giving users control.	– Implement granular consent mechanisms, allowing users to choose data-processing options.	– Allowing users to customize ad preferences and opt out of targeted advertising in an online platform.
	Respecting the autonomy and dignity of individuals regarding their data.	– Allow users to update and correct their personal information.	– Providing a self-service portal for users to edit their profile information on a social networking site.

4.5 PRIVACY BREACHES AND DATA LEAKS: ETHICAL IMPLICATIONS

Unfortunately, the incidence of privacy breaches and data leaks has become a reality in a connected society driven by data. These occurrences, which involve unlawful access to and disclosure of private information, have serious ethical ramifications for people, organizations, and society.

4.5.1 DATA BREACHES

Data breaches are a crucial intersection where morality, law, and social effects collide. Data breaches at their heart involve unlawful access to private information and offer a complex problem with important ethical implications. These breaches take many different forms, including hacker-initiated assaults, insider threats presented by employees inside businesses, and inadvertent exposes resulting from lax security procedures or setup errors.

Data breaches seriously infringe on people's right to privacy, according to ethical standards. They undercut people's faith in businesses to protect their private information. Such violations breach trust since they reveal sensitive information without permission. Importantly, data breaches cause harm to people, including identity theft, financial losses, emotional suffering, and reputational damage. When very sensitive information, such as personal health records, is compromised, the severity of the damage is increased.

Corporations and institutions in charge of gathering and maintaining personal data lose credibility due to data breaches. People may start to worry about the security of their data due to this trust gap, which has the potential to spread throughout whole businesses and the digital ecosystem. Governments and regulatory agencies respond by imposing legal and regulatory penalties, such as fines and legal action, on businesses that do not sufficiently secure personal information.

Organizations have a clear ethical duty to navigate the murky waters of data breaches. This duty includes making significant investments in cybersecurity defenses, quickly alerting those impacted after a breach, and taking preventative actions to minimize the damage. Thus, understanding data breaches requires an awareness of technological security measures and the ethical responsibilities associated with maintaining personal information.

4.5.2 IMPACT OF DATA BREACHES ON INDIVIDUALS

Data breaches have serious, morally important effects on people. Individuals are at risk for a variety of negative outcomes due to data breaches, including identity theft, financial loss, emotional suffering, and reputational harm. When personal information is compromised, there is a direct danger of identity theft, which may result in fraudulent acts and serious financial and legal implications. Another ethical issue is emotional anguish since victims of data breaches often feel anxious, afraid, and violated. Disclosing private or humiliating material may also damage one's reputation, hurting one's interpersonal connections and career prospects.

Since stolen data may be sold and utilized in fraudulent schemes, data breaches make people more susceptible to future cyberattacks. Since personal information is supposed to be kept private and shared only with permission, there has been a fundamental breakdown of confidence in privacy. Long-term effects may also persist, such as the persistent concern about the theft of medical information due to a data breach. Understanding the moral ramifications of these effects highlights the need for corporations to use strict data protection procedures, be transparent, and be accountable. In order to maintain confidence in the digital era, ethical reactions to data breaches must include prompt reporting, assistance for those impacted, and measures to minimize damage.

4.5.3 ACCOUNTABILITY AND RESPONSIBILITY

The harsh reality of data breaches in the digital age raises important moral questions about duty and accountability. Organizations that gather and handle personal data must understand their stewardship of people's private information. Strong security measures, risk assessments, and adherence to regulatory requirements, such as GDPR-compliant data protection legislation, are all part of this ethical commitment. Transparency is necessary for ethical responsibility; firms must swiftly and honestly notify impacted parties and regulatory authorities of breaches to create confidence even during a crisis.

Ethical responsibility is brought to the fore when an organizational failure results in a data breach. Negligence may result in breaches in security procedures, such as failing to patch urgent updates or disregarding known vulnerabilities. Ethical responsibility means admitting these errors, reflecting on them, and making amends to avoid such mistakes in the future. Organizations must also assist those harmed, including identity theft protection and credit-monitoring services, to uphold their moral and legal commitments.

Data breaches provide a singular ethical opportunity for development. These instances are seen as important teaching opportunities by ethical organizations, who use them to improve their data protection procedures, educating more people about cybersecurity via the dissemination of these courses.

4.5.4 ETHICAL RESPONSE TO DATA BREACHES

Data breaches are not only security events but also ethical problems that call for responsible and open solutions. The first step in an ethical reaction to a data breach is immediate communication between the impacted parties, government agencies, and the general public. Transparency is essential because it fosters confidence and enables people to take the appropriate safeguards by being open about the severity of the breach and any possible consequences. Support for persons affected by the breach is a top priority for ethical organizations, and they provide services like credit monitoring and financial loss compensation. Ethical approaches see data breaches as teaching moments. Organizations carry out rigorous post-breach investigations to find the reasons and vulnerabilities of breaches. As a result of this data, data security mechanisms are improved, policies are changed, and employees are trained to stop more breaches. Cooperation between ethical groups and regulatory and law enforcement bodies promotes accountability and upholds the rule of law.

Organizations often conduct press conferences or release statements to regain the public's faith. Additionally, it interacts with all its stakeholders—clients, staff, and shareholders—to manage expectations and uphold openness. Importantly, ethical responses refrain from blaming the victim and concentrate on the steps organizations may take to avoid such violations in the future.

4.6 ETHICAL APPROACHES TO PRIVACY MANAGEMENT IN ORGANIZATIONS

Organizations have substantial ethical issues maintaining privacy in a world where data has become a valuable commodity and data breaches are becoming more frequent. Privacy management is a core ethical duty and a question of compliance with laws. Companies that follow moral guidelines while managing privacy show their stakeholders that they care about upholding people's rights and establishing trust.

4.6.1 PRIVACY GOVERNANCE AND COMPLIANCE

Privacy governance and compliance are the foundations of ethical data management. These procedures include the structure, guidelines, and procedures businesses set up to guarantee the ethical and responsible handling of sensitive data while abiding by the law. The appointment of a Chief Privacy Officer (CPO) or DPO to manage compliance activities is a crucial aspect of privacy governance [6, 7]. This function includes monitoring compliance with data protection laws, doing risk analyses to stop breaches, and creating rules that specify ethical data processing procedures.

Ethical companies support the idea of "privacy by design." This proactive strategy incorporates privacy concerns from the beginning of creating products and services. This approach includes data reduction, security safeguards, and clear user permission processes. Beyond simple compliance, moral organizations promote privacy preservation as a moral obligation and acknowledge privacy as a basic right.

4.6.2 EMPLOYEE PRIVACY RIGHTS

Any firm has a basic ethical duty to respect the privacy rights of its employees. When managing employee data, ethical firms place a high priority on transparency. It is crucial to provide unambiguous information about the sorts of data gathered, the uses for which they will be used, and the duration of their retention. Employees must be completely aware of how their data is handled inside the company. When appropriate, ethical companies ask for employee permission before collecting and processing personal data. Employees are given enough information to make educated choices and can decline without repercussions. Thus, this consent is informed. Privacy and employee monitoring must be balanced. Even though monitoring is sometimes required for security reasons, ethical firms find a balance by creating explicit monitoring procedures, ensuring measures are fair and appropriate, and warning workers. Additionally, they provide systems for protecting whistleblowers so that workers may disclose privacy violations and other wrongdoing in a secure setting.

4.6.3   BUILDING A CULTURE OF PRIVACY

Organizations must prioritize privacy as a cultural value and a legal need. Building a privacy-conscious culture is essential to ethical privacy management because it ensures that organizational values like privacy are recognized and ingrained. Ethical organizations give education and awareness top priority as essential components. They know that an educated staff can better identify and reduce privacy threats. Employees learn about privacy concepts, laws, and particular company rules via frequent training sessions and awareness efforts. Leadership is crucial in this culture; moral leaders support and provide an example of privacy-conscious conduct. When senior leadership stands out for privacy, it makes it very apparent that ethical data processing is a given. By allocating funds, these leaders show their steadfast support for privacy projects.

Ethical businesses allow staff to express privacy concerns without fear of retaliation. Transparent means for reporting potential privacy violations strengthen the organization's commitment to moral data management. The cornerstone is accountability, ensuring that people and teams are held accountable for following privacy regulations. A culture of privacy is also firmly ingrained in everyday actions and not just policy texts. By design, ethical firms prioritize privacy and incorporate privacy principles into every aspect of their business. This includes user-centered consent procedures, strong security safeguards, and data minimization.

A culture of privacy is dynamic rather than stagnant. Organizations with a strong ethical foundation strive for constant development. They continuously evaluate and update privacy policies, procedures, and training to keep current with new privacy concerns and ethical considerations. This flexibility ensures that the company is robust and responsive to the changing technological and regulatory context.

REFERENCES

1. Yilma, K., *Privacy and the role of international law in the digital age.* 2022: Oxford University Press.
2. Hoofnagle, C.J., B. Van Der Sloot, and F.Z. Borgesius, The European Union general data protection regulation: What it is and what it means. *Information & Communications Technology Law*, 2019. 28(1): pp. 65–98.
3. Mulgund, P., Mulgund, B. P., Sharman, R., & Singh, R, The implications of the California Consumer Privacy Act (CCPA) on healthcare organizations: Lessons learned from early compliance experiences. *Health Policy and Technology*, 2021. 10(3): p. 100543.
4. Wright, D., Enforcing privacy, in David Wright and Paul De Hert (eds) *Enforcing* privacy*: Regulatory,* legal *and* technological approaches. 2016, Springer. pp. 13–49. 5. Fritsch, C., Data processing in employment relations; Impacts of the European general data protection regulation focusing on the data protection officer at the worksite, in *Reforming European Data Protection Law*. 2014, Springer. pp. 147–167.
6. Bantan, M. and M. Shawosh, *Chief privacy officers: A literature review.* 2021: AMCIS.
7. Lyons, V. and T. Fitzgerald, *The privacy leader compass: A comprehensive business-oriented roadmap for building and leading practical privacy programs.* 2023: CRC Press.

5 Freedom of Expression
Ethical Boundaries in the Digital Realm

5.1 UNDERSTANDING FREEDOM OF EXPRESSION IN THE DIGITAL AGE

There have been major changes to the idea of freedom of speech. Understanding freedom of speech in the digital era requires understanding its cultural background, complicated legal issues, and the ever-changing internet. Our understanding of balancing and preserving this basic freedom and the moral obligations connected with online communication needs to advance along with technology as it does.

5.1.1 THE EVOLUTION OF FREEDOM OF EXPRESSION IN THE DIGITAL ERA

The way people use their right to free speech has undergone a significant transition since the dawn of the digital age. In the context of internet and digital technology, this part investigates the dynamic growth of this essential democratic right. As guaranteed by international human rights treaties and constitutional laws, freedom of speech has long been a tenet of democracies. In the past, it has meant the freedom for people to speak their minds without worrying about being silenced or facing retaliation from the government. While this idea is constant, the digital era has brought new difficulties and possibilities.

How information is produced, shared, and consumed has undergone a fundamental transformation due to the digital revolution, characterized by the widespread use of the internet and digital technology. The internet has made it possible for almost anybody to become a content producer, publisher, or distributor, in contrast to conventional media, when a few people control the flow of information. The information flow has become more inclusive and democratic due to this empowerment.

The accessibility and worldwide reach of information are two of the most important features of the digital age. Because the internet has no geographical boundaries, people from all over the globe may interact and exchange ideas. This has aided intercultural communication and given voice to minority and disadvantaged ideas. Additionally, technology has made it possible for information to circulate quickly, positively, and negatively, influencing the worldwide public conversation.

Because of digital tools and platforms, people now have more power than ever to engage in public debate. People may communicate their ideas and views via social media, blogs, podcasts, and video-sharing websites. As a result, grassroots action, citizen journalism, and online groups based on similar interests and ideals have all grown.

 DOI: 10.1201/9781003584452-5

The digital age has created new platforms for expression, but it has also brought forth issues with information overload. Because there is a wealth of information accessible online, people need to wade through a sea of information to identify reputable and trustworthy sources. Because so much information is available, it may sometimes result in the spread of false information and a decline in faith in established institutions. Ethical issues related to freedom of speech are becoming more crucial as the digital age develops. Questions of cyberbullying, hate speech, misinformation, and the duties of internet platforms in controlling material have gained attention [1]. It still needs to be easier to balance defending free expression and addressing these moral issues.

5.1.2   LEGAL FRAMEWORKS AND ONLINE SPEECH

Due to the internet's worldwide reach and the variety of ways people and organizations express themselves online, the legal framework governing freedom of speech has become more complicated. As a basic human right, freedom of speech is susceptible to varying standards and interpretations worldwide. To maintain public order and protect individual rights, some countries prioritize absolute free speech, enabling people to express themselves without intervention from the government.

Because the internet has no physical borders, it may be difficult to determine which legal systems govern online speech and information. Online material may be accessible from anywhere, which raises the issue of whose laws should apply: those of the nation where the information is hosted, the place where it is accessed, or the country where the content was created. Conflicting legal systems among many countries make things even more complicated, demanding diplomatic missions and international collaboration to resolve these jurisdictional problems.

Intermediary liability is a crucial component of the legal framework for online communication in the digital era. Many nations hold internet service providers, such as social networking sites and web hosts, accountable for the material that their users publish online. This strategy forces these middlemen to set content restriction measures to prevent legal penalties. Nevertheless, this technique raises questions about censorship, overreach, and the significant power that private organizations have over the kind of information allowed on their platforms.

It continues to be difficult to strike a balance between the preservation of free speech and the elimination of dangerous or unlawful information. The complexity of this legal environment is largely due to the changing role of internet intermediaries in defining the parameters of online speech, contradictions between national and international laws, and tensions between private regulation and public interest. In the digital era, finding an ethical balance for regulating online speech becomes more and more important as technology develops.

5.1.3   GLOBAL PERSPECTIVES ON DIGITAL FREEDOM OF EXPRESSION

The topic of digital freedom of speech is complex, with many different worldwide viewpoints that are impacted by cultural, political, and historical considerations. How nations negotiate this right is greatly influenced by cultural norms and beliefs. While some cultures prioritize social peace and censor harmful information, others support individual speech, especially when it contains divisive opinions. These viewpoints are further shaped by political ideologies, with democratic countries upholding free speech as a fundamental democratic virtue and authoritarian regimes favoring censorship and monitoring to retain control.

Despite these differences, widespread dangers to online freedom still exist. Many governments use censorship and monitoring to regulate the flow of information, often crushing opposition. Independent media organizations and journalists who criticize governments are subject to cyberattacks intended to silence them. Additionally, private tech businesses have a lot of influence on digital freedom because of their content-filtering practices, which has sparked discussions about balancing corporate accountability and individual liberties.

In the face of these difficulties, advocacy and activity are essential. Civil society organizations, human rights organizations, and proponents of free speech put forth much effort to raise awareness about risks to online expression and lobby for laws that protect it. Often in tremendous personal danger, journalists and activists make sure that knowledge is freely shared. International human rights laws provide a foundation for defending digital freedom of speech globally, but their efficacy depends on how willingly countries support and implement these laws.

5.2 ETHICAL CHALLENGES OF BALANCING FREE SPEECH AND HATE SPEECH

The fragile balance between free expression and hate speech is one of the most urgent and difficult ethical concerns. However, the internet and social media development has significantly accelerated the spread of intolerance, prejudice, and discrimination. Expressions that inspire violence, prejudice, or animosity against people or groups because of characteristics like race, religion, ethnicity, gender, sexual orientation, or disability are considered hate speech. Despite the widespread condemnation of hate speech, many countries struggle to deal with it without violating the revered ideal of free speech.

5.2.1 DEFINING HATE SPEECH IN THE DIGITAL CONTEXT

In the contemporary era of communication, defining hate speech in the internet setting is a difficult and ever-evolving task. A comprehensive understanding of what constitutes hate speech is required to successfully balance the need to counteract harmful speech online and preserve free expression in light of the fast proliferation of digital platforms. Language is highly flexible, and context is very important in the digital world. Depending on the situation, innocent words or phrases may have very negative or hurtful connotations. It is important to distinguish between hate speech and offensive speech; essential markers include discrimination, encouragement of violence, or the dehumanization of certain groups.

There are differences in the concept of hate speech between cultures and legal systems. The global nature of the internet and the need to address hate speech within a multicultural setting are highlighted by the fact that what is deemed hate speech in one jurisdiction may not fulfill the same requirements in another. Digital platforms play a crucial role in influencing the dialogue by setting community norms to control hate speech. However, these requirements might differ significantly across platforms, which causes inconsistent enforcement. Another difficulty level is added by employing algorithms for content-filtering since automated tools may need help identifying hate speech owing to linguistic subtleties, contextual clues, and cultural considerations.

5.2.2   The Impact of Hate Speech on Online Communities

Online groups are under the sway of hate speech, a problem that is ubiquitous in the digital sphere and has far-reaching effects. It is a formidable barrier to free speech and open debate, often motivating people to self-censor to avoid becoming the objects of abuse. This chilling effect reduces the variety of voices in online forums, stifles the depth of dialogue, and eventually silences crucial viewpoints.

However, hate speech has damaging implications that go beyond simple self-censorship. It encourages poisonous online communities and creates a tone of division that normalizes racist conduct. Such negativity degrades the quality of online interactions and undermines productive conversation and teamwork. Hate speech has a significant negative psychological and emotional impact on those who are exposed to it, leaving them with feelings of worry, tension, melancholy, and loneliness. One's sense of self-worth and belonging in digital settings is undermined by this ongoing exposure to dehumanizing words, continuing a cycle of damage.

The implications of hate speech can include a decline in community trust online. When hate speech is allowed to flourish, users may lose trust in the communities and platforms they interact with, which may reduce user interaction and cause online spaces to be abandoned. In its most severe forms, hate speech has the potential to cause physical damage, aiding in the radicalization of individuals online and encouraging violence against certain communities. Additionally, hate speech imperils the inclusion and diversity principles that support vibrant online communities by silencing minority voices and rejecting certain viewpoints. This prevents advancement and innovation and deprives these places of the collective knowledge they may provide. The urgent need for moral strategies to lessen the impacts of hate speech, such as effective moderation, community norms, and an environment of respect and empathy online, is highlighted by awareness of the complex repercussions of hate speech. Additionally, addressing the underlying factors that contribute to hate speech, such as prejudice and discrimination, is crucial for developing a more equal and peaceful digital environment.

5.2.3   The Challenge of Regulating Hate Speech Ethically

The ethical control of hate speech is difficult in the digital era and requires careful balancing. Hate speech, defined as statements that encourage violence, prejudice, or animosity toward people or groups based on various characteristics, presents difficult moral decisions for society, governments, and internet platforms. It is a difficult and diverse task to address this issue while preserving the values of free speech and individual rights.

The distribution of culpability is one of the first moral conundrums in regulating hate speech. Should governments or private internet platforms have a part in it? Should not both work together? Each strategy has a unique set of ethical issues. While private platform moderation raises concerns about the concentration of power and transparency, government regulation runs the possibility of censorship and the suppression of dissident views. Private organizations often decide what constitutes hate speech and how it should be dealt with, raising questions about prejudice and limiting public accountability.

The definition of hate speech in the context of the internet is at the core of the ethical dilemma of controlling hate speech. Hate speech is, by its very nature, subjective and situational. One person's definition of "hate speech" may not match another person's definition of "legitimate expression of opinion." It is a constant battle to strike the correct balance between suppressing hate speech and upholding the right to free expression. Collaborative measures, including governments, internet platforms, civil society, and people, are crucial to combat the detrimental effects of hate speech while upholding the core values of free speech and open conversation in the digital sphere. Since the digital ecosystem is always changing and adding new complications, finding ethical answers to this problem is never-ending.

5.3   ONLINE DISINFORMATION AND THE SPREAD OF FAKE NEWS

News and information are being disseminated at an unprecedented rate in the era of digital technology. While this has given people access to a plethora of knowledge and empowered them, it has also given birth to a worrying phenomenon: the proliferation of false news and online misinformation. This section delves into the complicated world of false information, disinformation, and fake news while examining these problems' moral ramifications.

5.3.1   The Proliferation of Disinformation Online

The spread of false information online has seriously threatened the integrity of information ecosystems and the moral bounds of free speech. The purposeful dissemination of incorrect or misleading information with the aim of deceiving is known as disinformation, and the internet has become a favorable environment for it. Several variables have caused its quick development. First, anybody with access to the internet and the ability to publish material can produce and spread information, making it simple for individuals and organized organizations to launch misinformation operations. Second, social media platforms, which encourage user interaction, often unwittingly propagate misinformation by promoting dramatic or emotionally charged material, regardless of its authenticity. Third, it might be difficult to track down and hold people propagating misleading information responsible since online anonymity and pseudonymity act as a shield for them.

The dissemination of false information online has far-reaching effects. Inaccurate information may harm reputations, skew public debate, sway elections, inspire violence, and undermine faith in dependable information sources. It is critical to understand that combating misinformation does not need the suppression of free speech; rather, it necessitates a sophisticated strategy that strikes a balance between the right to free speech and the moral obligation to stop the destructive spread of false information. Promoting media literacy, encouraging fact-checking, and encouraging ethical online activity should be the main goals of campaigns to fight misinformation. These tactics, driven by ethical concerns, may aid in protecting the essential democratic society's ideal of free speech.

Ethical concerns must guide efforts to address the challenges posed by the spread of internet misinformation, as society seeks to balance the protection of free speech with the need to curb harmful rumors. Achieving this balance requires a cultural shift toward critical thinking, responsible information consumption, and the implementation of technological solutions and legislative measures.

5.3.2 COGNITIVE BIASES AND DISINFORMATION CONSUMPTION

These biases are mental heuristics that often cause people to make deliberate mistakes in judgment and choice. Confirmation bias is one of the most prevalent biases, wherein individuals recall and seek out information supporting their opinions, fostering echo chambers and filter bubbles. These prejudices are amplified even more by algorithms and social networks, which foster conditions where people are constantly exposed to data that supports their preconceptions.

The illusory truth effect makes the problem worse, which increases people's propensity to accept incorrect information with repeated exposure. Furthermore, misinformation may take advantage of the anchoring bias by using a false or misleading title as an anchor to skew views. On the other hand, the Dunning-Kruger effect may cause people to confidently spread false information without fully appreciating its accuracy or repercussions [2, 3].

Fighting the spread of false information requires addressing certain cognitive biases. People may be equipped to notice and counteract these biases by being encouraged to develop their critical thinking and media literacy abilities. Modifications to algorithm design and online platforms may prioritize many viewpoints and reliable information, minimizing the influence of cognitive biases on the consumption of misinformation. Table 5.1 lists key cognitive biases, real-world examples of each bias, how it affects the intake of misinformation, and mitigation techniques for its effects on online information consumption.

5.3.3 ETHICAL APPROACHES TO COMBATING DISINFORMATION

Promoting media literacy and critical thinking is fundamental in this complicated environment. A society that is less vulnerable to manipulation is one where people are given the tools to assess sources, cross-reference data, and spot false material. The preservation of moral values is fundamentally dependent on responsible media. Accuracy and verification are priorities for news organizations, and fact-checking programs are crucial instruments for disproving misleading assertions. It is essential that different parties, including governmental institutions, tech companies, and civil society groups, work together. Together, they can create technological solutions that identify and counteract misinformation while upholding the values of accountability and openness.

The difficult job for legislators is to create legislation that tackles misinformation without restricting the freedom of expression. Campaigns for public awareness that stress the value of fact-checking and ethical information sharing may educate people about the dangers of misinformation. Since misinformation often crosses national boundaries, international cooperation improves attempts to counteract it. The ethical fight against misinformation is a multidimensional team effort that protects free speech while safeguarding the accuracy of information in the digital era.

TABLE 5.1
Cognitive biases in disinformation consumption: Examples and mitigation strategies

Cognitive Bias	Example Manifestations	Impact on Disinformation Consumption	Mitigation/Addressing Strategies
Confirmation Bias	– Only reading news articles that align with one's political beliefs.	Reinforces echo chambers and filter bubbles, facilitating the spread of disinformation.	Encourage diverse news sources and promote critical thinking and fact-checking to counteract confirmation bias.
Filter Bubbles & Echo Chambers	– Seeing primarily posts from like-minded friends on social media.	Isolates individuals within their information bubbles, limiting exposure to diverse perspectives.	Design algorithms that prioritize diverse viewpoints and encourage cross-discussion platforms.
Illusory Truth Effect	– Believing a false health claim after seeing it in multiple online forums.	Increases the perceived credibility of disinformation through repetition.	Promote media literacy and fact-checking, emphasizing the importance of verifying information from multiple sources.
Anchoring Bias	– Accepting the initial headline as fact, even when later evidence contradicts it.	False or misleading information can serve as an anchor, influencing perceptions and judgments.	Encourage individuals to seek additional information and consider multiple perspectives before forming conclusions.
Dunning–Kruger Effect	– Sharing a medical claim without expertise or evidence, assuming it is accurate.	This can contribute to the spread of disinformation by individuals who need to be made aware of their lack of expertise.	Promote self-awareness and humility, encourage individuals to consult experts, and provide accessible fact-checking resources.

5.4 ETHICAL IMPLICATIONS OF CONTENT MODERATION AND CENSORSHIP

Having the authority to control and censor information on internet networks has become essential and often divisive. Although content moderation aims to preserve a courteous and secure online environment, it poses serious ethical questions about free speech, transparency, and the influence of technology firms on public conversation.

5.4.1   THE ROLE OF PLATFORMS IN CONTENT MODERATION

Online forums have become the main exchange points for knowledge and public dialogue. These platforms, which include social networking sites, websites for sharing material, and discussion boards, act as online town squares where people can express their opinions, interact, and discuss a wide range of subjects. The necessity to control and govern the material circulating via these digital spaces has become an urgent concern due to the spike in user-generated content.

Online platforms greatly impact the kind and quality of material accessible to their sizable user populations. They create community standards and content regulations that outline what constitutes appropriate conduct and material on their platforms. These rules cover various topics, including graphic material, disinformation, and hate speech. Content moderation is largely used to enforce these rules.

Monitoring, assessing, and, where required, deleting or limiting user-generated material that violates platform regulations constitutes content moderation. It uses automatic tools, such as algorithms that alert users to potentially harmful information, and human moderators who assess each case individually and make choices. The goal of content moderation is to establish and maintain a safe, civil, and welcoming online space where users may express themselves without fear of abuse or violence.

However, there are certain difficulties with content moderation. Defining what constitutes appropriate material may be difficult since various cultures, groups, and people may have different ideas about what is insulting or damaging. Additionally, the enormous amount of information produced on these platforms calls for a greater dependence on automated systems, which have problems, including algorithmic bias and possible mistakes. Transparency and accountability have grown to be crucial among these difficulties. In order to ensure that users and the general public understand how these systems work, how information is identified or deleted, and how appeals may be made, platforms are being encouraged to give more transparent insights into their content moderation procedures and choices.

Platforms' involvement in content regulation requires them to strike a difficult balance. Platforms need to safeguard the right to free speech while avoiding unjustified censorship. On the other hand, people are responsible for protecting themselves from the damage that certain material kinds, such as harassment and hate speech, might cause. This balance needs to be maintained via constant ethical reflection, strong regulations, and, sometimes, unpleasant choices. Platform involvement in content regulation determines how individuals interact with digital environments daily. Platforms need to navigate moral conundrums and difficulties as custodians of these online communities, working toward the common objective of establishing positive, welcoming, and courteous online communities that will significantly impact how people communicate and interact in the future.

5.4.2   CHALLENGES IN DEFINING AND ENFORCING COMMUNITY STANDARDS

Defining and upholding community standards on online platforms is always changing and fraught with enormous obstacles. The first significant issue is caused by online groups' many cultural and contextual variances. Establishing universally applicable rules is challenging since cultural norms for what is offensive or acceptable vary greatly. Platforms need to negotiate these complexities while developing standards that consider different tastes.

The second problem is that the internet is always changing. Community standards often need to be updated to address new challenges as norms, language, and expressions change over time. Preserving the essential value of speech freedom must strike a balance with this progress. The final difficulty is finding this equilibrium as platforms struggle with the moral conflict of upholding free expression while avoiding damage. A persistent problem arises from the range between more restrictive standards that restrict speech and laxer regulations that permit harmful information.

The fourth problem is the size of content moderation, which is enormous on major platforms with billions of users and a constant stream of information. At this size, enforcing community standards needs both human moderation teams and computers, yet this often results in mistakes and inconsistent decision-making. The use of algorithms creates the fifth problem, known as algorithmic bias. These artificial intelligence (AI) systems may unintentionally bias against certain groups and have difficulty understanding subtle context, leading to improper content removal or retention.

The general public and users expect detailed justifications for content moderation decisions and channels for appeals. Platforms also need to navigate a variety of international legal systems and legislation. Last but not least, figuring out whether user-generated material is intended as satire, comedy, or a real danger is a difficult task that may result in false positives and false negatives in content moderation. Table 5.2 concisely summarizes the difficulties in developing and enforcing community standards on online platforms, their effect, instances from the actual world, and mitigating measures.

5.4.3 BALANCING THE RIGHT TO FREE SPEECH WITH CONTENT REGULATION

A difficult ethical conundrum is the conflict between upholding the basic right to free speech and the need for content restriction. Democratic societies are built on the principle of free speech, which enables people to express their opinions without worrying about censorship from the government. However, there are particular difficulties in the internet environment. The worldwide reach of the internet magnifies both the distribution of destructive materials like hate speech and false information, which may have real-world repercussions and the interchange of various views.

The problem is determining the boundaries of free expression in the digital sphere. There are concerns concerning what constitutes encouragement to violence or hate speech, as well as when false information becomes detrimental. Ethical content regulation aims to find a careful balance between preventing damage to people and sustaining a lively flow of ideas.

Since too broad or biased content regulations may hinder legitimate speech and suppress underrepresented voices, censorship is a major issue in this discussion. To reduce these worries, accountability and transparency are essential. Users and the general public should have access to clear, uniformly applicable standards for content moderation and information on the function of human moderators and the appeals procedure.

Increasing digital and media literacy among users is another step in solving this ethical problem. By enabling people to evaluate online information critically, we can lessen our dependence on platform-level content management and better separate trustworthy content from false or harmful information. Technology firms, governments, civic society, and users need to work together to resolve the continuing discussion about balancing the right to free expression and content restriction.

TABLE 5.2
Difficulties in defining and implementing community standards

Challenges	Impact	Examples	Mitigation
Cultural and Contextual Variations	Inconsistent moderation. User alienation.	Cultural slang interpretation.	Diverse advisory input. Context-aware moderation.
Evolving Norms and Language	Outdated standards. User needs to be clarified.	Changing internet slang. Viral trends.	Regular updates. Clear communication.
Balancing Freedom of Speech and Harm Prevention	Restricting speech. Permitting hate speech.	Political discourse vs. hate speech. Satire vs. misinformation.	Expert consultation. Transparent policies.
Content Moderation Scale	Inconsistency. Errors.	Millions of daily posts. Countless video uploads.	AI screening. Human moderators.
Algorithmic Bias	Discrimination risk. Intent ambiguity.	Biased algorithms. Sarcasm misinterpretation.	Bias audits. Human oversight.
Transparency and Accountability	Trust erosion. Confusion.	Content removal queries. Opacity concerns.	Clear guidelines. Appeals. Communication.
Legal and Regulatory Variability	Compliance challenges. Policy conflicts.	GDPR, hate speech laws.	Legal teams. Localized policies.
Determining Intent and Context	Unfair removals. Nuanced cases.	Satire vs. hate speech. Humor vs. harassment.	Context AI. User input. Human moderators.

5.5   ONLINE HARASSMENT AND CYBERBULLYING

5.5.1   THE PSYCHOLOGICAL IMPACT ON VICTIMS

Cyberbullying and online harassment have a significant psychological effect on the victims, with potentially serious and long-lasting repercussions. The continual barrage of threatening messages and threats interrupts their everyday lives. It puts them in a constant state of tension, which causes the mental discomfort and anxiety experienced by those targeted to be overpowering. This ongoing concern may make victims retreat from both offline and online social connections, aggravating their mental distress and sometimes resulting in despair or even suicidal thoughts.

Online abuse victims can experience continuous anxiety, hypervigilance, and intrusive thoughts connected to the abuse, which can be so painful that it mimics post-traumatic stress disorder (PTSD) [4]. In addition to the emotional toll, insults, and character assassinations cause victims to lose confidence in themselves and their abilities. Nevertheless, some victims show resiliency by looking for assistance from friends, family, or online groups that support those who have been victimized by cyberbullying. Understanding the severe psychological effects of online harassment is essential for creating practical solutions to help victims and deal with this urgent problem.

5.5.2 ETHICAL RESPONSES AND STRATEGIES FOR PREVENTION

Online harassment and cyberbullying offer significant issues that need moral responses and preventative measures that uphold freedom of speech while addressing damage (Figure 5.1). To achieve this equilibrium successfully, a complex strategy is needed. First and foremost, campaigns for education and awareness are crucial. Online harassment and cyberbullying may be recognized and handled, especially by the younger generation, thanks to digital literacy training. Promoting media literacy also allows consumers to distinguish reliable sources from deceptive or dangerous information, improving their online experience.

Developing effective reporting methods on internet platforms is a crucial component of ethical responses. These programs should protect complainants' privacy while being transparent, effective, and responsive to user complaints of harassment or bullying. Options for anonymous reporting might be crucial for persons who worry about reprisals. Platforms should also implement clear community rules that forbid harassment and cyberbullying while upholding the right to free speech. Effective content moderation mechanisms are essential to find a balance between safeguarding users from damage and supporting their freedom to voice viewpoints, even controversial ones.

A fundamentally ethical reaction to internet abuse is support for the victims. Recognition of the emotional toll that harassment may have on people leads to the provision of mental health services and assistance. Additionally, giving victims who want to use the legal system to seek reparation guarantees access to justice. In order to create and put into practice moral solutions to prevent harassment, industry players, governments, civil society groups, and users need to cooperate. It is crucial to encourage research into the causes and effects of online harassment and to share knowledge and best practices across platforms and stakeholders. A complete strategy to prevent online harassment and cyberbullying should include ethical bystander involvement, legislation that targets online harassment while upholding basic rights, and encouraging bystanders to assist victims or report incidents.

FIGURE 5.1 Ethical responses and strategies for prevention

5.6 PROTECTING FREE SPEECH WHILE UPHOLDING ETHICAL STANDARDS

Preserving free speech is crucial to maintaining a healthy and democratic society in an increasingly linked world where information travels freely via digital means. This freedom needs to be tempered with moral guidelines that guard against damage and preserve online civility since it is not unrestricted.

An essential paradigm for managing the complicated web of online speech and dialogue has emerged: a "free but responsible" internet. It acknowledges the moral obligations that accompany maintaining the values of free speech and open communication while also recognizing the basic necessity of doing so. In order to foster innovation, education, and democratic engagement, a "free but responsible" internet emphasizes that the internet should continue to be a platform for the unrestricted exchange of ideas. This will allow people to express themselves, share information, and participate in public discourse without undue censorship or government interference.

This independence has limitations, however. It entails several duties that people, online groups, and governments need to take care of. These obligations include upholding the rights and dignity of others, making sure that information shared online is accurate, protecting user privacy and personal information, implementing responsible content moderation techniques, encouraging digital literacy, encouraging critical thinking, and creating and enforcing legal frameworks that uphold free speech rights while addressing the negative effects of online behavior.

The idea of a "free but responsible" internet acknowledges that unfiltered internet use may have unfavorable effects, including the propagation of hate speech, false information, and cyberbullying. We need to work together to establish a digital environment that encourages both freedom and responsibility. In order to achieve a "free but responsible" internet, a careful balance needs to be struck between upholding the ethical norms that protect everyone's well-being and dignity online and the ideals of free speech.

5.6.1 Technological Solutions for Ethical Content Management

Internet information management has become a serious problem that calls for cutting-edge technology solutions that adhere to moral principles. Among these options, content-filtering algorithms supported by AI and machine learning stand out as essential instruments for automatically locating and reporting information that contravenes moral or legal standards. However, the possibility of false positives and negatives presents an ethical conundrum as these algorithms need to traverse a massive sea of material.

User reporting tools allow users to report inappropriate content and are also essential to ethical content management. These solutions encourage user participation in moderation but also pose ethical questions about openness, abuse prevention, and upholding the integrity of the reporting process. It is essential to provide users with information about why certain material was deleted or marked to increase the moderating process's openness and accountability.

Content rating systems let individuals customize their material intake depending on their beliefs and tastes; however, worries about the possibility of filter bubbles still exist. Collaborative moderation distributes the burden of upholding content standards by including user communities in the process of jointly evaluating and reporting information. However, this strategy calls into question the need for supervision and community prejudice. Sentiment analysis technologies powered by AI that can detect emotional tones and context are essential for spotting minor instances of harassment or hate speech. However, the ethical issues that need to be overcome include accuracy and the danger of overreaching content labeling. Maintaining free expression and ethical norms in the digital sphere is still challenging, calling for ongoing technical advancement to handle new ethical issues and foster a welcoming and secure online community.

5.6.2 COLLABORATIVE APPROACHES TO SAFEGUARDING FREEDOM OF EXPRESSION

Protecting freedom of speech while respecting ethical norms has become a complex task in an age characterized by the fast growth of digital information and the world-wide accessibility of internet platforms. Given the difficulty of this undertaking, collaborative methods have become a potential answer. Collaboration is crucial because only some organizations, including governments, tech corporations, or civil society, can handle all the complex problems of online freedom of speech. These challenges include a broad range of difficulties, including hate speech, misinformation, harassment, and censorship, all of which need a variety of viewpoints and specializations to be resolved effectively.

Multi-stakeholder alliances are at the forefront of coordinated initiatives to safeguard online freedom of speech. Various parties, including governmental entities, IT firms, non-governmental organizations (NGOs), academic institutions, and online communities, are involved in these relationships. Multi-stakeholder initiatives strive to balance defending free expression and tackling harmful internet material by bringing together this varied group of people.

Collaboration techniques emphasize shared accountability and responsibility. For instance, tech corporations are becoming more aware of their responsibility to reduce the harmful effects of their platforms. They collaborate with other organizations, experts, and users to create content moderation rules and practices that align with societal norms and ethical standards. These cooperative endeavors' fundamental tenets include transparency, independent ethical monitoring, worldwide collaboration, and user empowerment.

REFERENCES

1. Arora, A., Nakov, P., Hardalov, M., Sarwar, S.M., Nayak, V., Dinkov, Y., Zlatkova, D., Dent, K., Bhatawdekar, A., Bouchard, G. and Augenstein, I., Detecting harmful content on online platforms: What platforms need vs. where research efforts go. *ACM Computing Surveys*, 2023. 56(3): pp. 1–17.
2. Mugg, J. and M.A. Khalidi, Self-reflexive cognitive bias. *European Journal for Philosophy of Science*, 2021. 11(3): p. 88.
3. Canady, B.E. and M. Larzo, Overconfidence in managing health concerns: The Dunning–Kruger effect and health literacy. *Journal of Clinical Psychology in Medical Settings*, 2023. 30(2): pp. 460–468.
4. Ahmed, N., *Cyberstalking: A content analysis of online gender-based offenses*. 2019.

6 Intellectual Property
Ethical Considerations in the Digital Landscape

6.1 THE CONCEPT OF INTELLECTUAL PROPERTY IN THE INFORMATION AGE

Intellectual property (IP) has experienced a significant shift that has broadened its use and value. Fundamentally, IP refers to the legal protections afforded to works of art, inventions, designs, and more. The definition of IP has expanded in the Information Age to include a wide range of digital assets and intangible inventions. At the same time, it was formerly restricted to patents, copyrights, and trademarks. Digital material, software, algorithms, data, and cutting-edge licensing schemes like Creative Commons are all part of this transformation. However, along with these developments, moral conundrums have emerged, such as problems with internet piracy, patent trolling, and worries about digital rights management.

The world of digital material is a crucial aspect of IP in the Information Age. Digital information, including articles, photographs, movies, music, and software, has become more widely available and shared as a result of the expansion of the internet. It is now necessary to have a comprehensive knowledge of fair use and ethical sharing since legal frameworks have changed to manage these digital assets. Further development identifies software and algorithms as valuable IPs, raising moral concerns about using open-source vs. private software and responsible algorithms, particularly in artificial intelligence and machine learning.

The big data age has created ethical challenges regarding data ownership, privacy, and sharing. Both businesses and people must avoid these problems while upholding IP rights. Alternative license models, including Creative Commons, have evolved in reaction to this dynamic environment, allowing artists to choose how their work may be used [1]. However, moral conundrums continue, necessitating a careful balancing act between preserving IP rights and promoting innovation and information flow.

6.1.1 HISTORICAL EVOLUTION OF IP

IP has a rich, complex past that has evolved through time to satisfy society's and technology's changing requirements (Figure 6.1). Its roots may be seen in earlier civilizations when writers and innovators sometimes received monopolies or limited acknowledgment for their works. While on a limited scale, these early forms of IP attempted to promote creativity and original thought.

The basis for the patent system was formed during the medieval era in Europe by the role that guilds and craft groups played in preserving the know-how and methods

DOI: 10.1201/9781003584452-6

of their professions. They also protected artisans and inventors seeking exclusive rights for their inventions. Early copyright laws were created in Europe due to Johannes Gutenberg's development of the printing press in the 15th century. These laws shifted the emphasis from protecting physical copies to protecting intellectual inventions themselves.

More thorough IP protection was required due to the Industrial Revolution in the 18th and 19th centuries, which saw the emergence of new technologies in equipment, manufacturing techniques, and branding. The mechanisms for patents and trademarks were improved and enlarged to accommodate these advancements. In the late 19th and early 20th centuries, international accords and conventions such as the Berne and the Paris Convention set universal IP rules.

Digital technology has brought up new possibilities and difficulties for IP today. Questions concerning copyright enforcement and fair use have been raised in light of how simple it is to duplicate and distribute digital information. The IP environment has also changed due to concerns like software patents and open-source initiatives. As a result of the current quick pace of technological development and globalization, IP has expanded to include digital assets, data, and software. This development prompts continuing discussions about balancing preserving IP rights and fostering creativity and innovation in the Information Age.

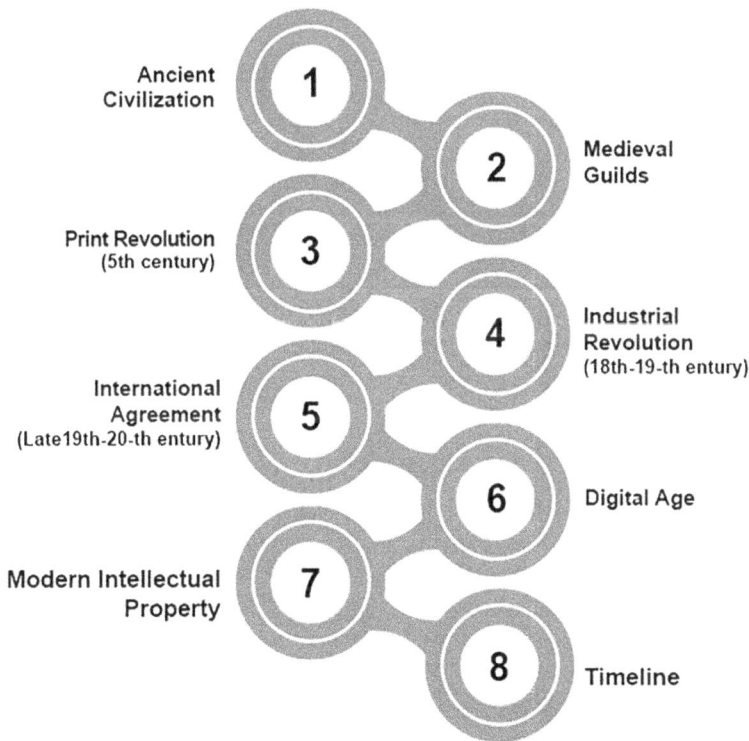

FIGURE 6.1   Evolution of IP: A simplified historical overview

6.1.2   IP in the Digital Age

The digital era has ushered in a new era for IP, redefining how we create, distribute, and safeguard intangible assets. In this context, IP encompasses a broad range of digital assets, including software, algorithms, digital media, and data. Understanding the complexities of IP in the digital era is indispensable for navigating the ethical and legal landscape.

Digital content is ubiquitous in the current digital age. Music, movies, literature, images, and software are now commonly distributed and ingested digitally. However, this transition has posed challenging obstacles in the domain of IP. Originally designed for physical media, copyright laws have had to be modified to address digital content, resulting in debates about piracy, fair use, and digital rights management (DRM) [2].

Previously niche disciplines, software, and algorithms are now integral to modern society. The complexities of IP rights in software include debates over open-source and proprietary software and the ethics of software development. Concerns about transparency, bias, and responsible use have been spurred by the proliferation of algorithms, particularly in disciplines such as artificial intelligence and machine learning. In addition, the data-driven nature of the digital age has spawned ethical debates regarding data ownership, privacy, and responsible data sharing, which affect both individuals and organizations.

6.2   COPYRIGHT, FAIR USE, AND ETHICAL DECISION-MAKING

The fundamental principle of IP law, known as copyright, was created to safeguard both people's and institutions' creative and original works. The bounds of copyright, however, have become more convoluted in the digital era, posing significant ethical issues regarding how to strike a balance between the rights of artists and the public's ability to access and utilize information. Exclusive ownership of creative works, such as literary, artistic, and musical compositions, is granted to creators under copyright law. This protection promotes innovation by offering artists a financial incentive to make and distribute their work. However, there are restrictions and exceptions to copyright protection.

6.2.1   Fair Use Doctrine and Its Limits

The fair use doctrine, a tenet of copyright law, is an adaptable rule intended to encourage the ethical use of protected content without requiring express consent from copyright holders. It is a crucial protection for the right to free speech and the spread of information. Courts look at four important considerations when determining whether a use falls within the definition of fair use: the user's intent and character, the nature of the copyrighted work, the size and significance of the portion utilized, and the impact on the original work's market value.

Fair use has its limitations, however. It does not authorize the wholesale duplication of copyrighted content for commercial reasons, but it does allow transformative uses that enhance or provide alternative viewpoints on original works. The concept seeks to achieve a compromise between protecting the rights of authors and promoting a strong public knowledge base. However, other elements and situations could be considered because it is still a case-specific judgment. This highlights the need for care and, if required, legal advice.

6.2.2 ETHICAL DILEMMAS IN COPYRIGHT ENFORCEMENT

The ethical complexities of copyright enforcement in the digital age pose problems for copyright holders and the general public. The possibility of aggressive enforcement by copyright holders is a serious worry. Even though copyrighted content may fall within the parameters of fair use, some may adopt aggressive strategies to safeguard IP, such as stop-and-desist letters or legal action. This overzealousness raises concerns about the moral balance between defending rights and protecting freedoms since it might inhibit genuine creative expression, restrict access to information, and discourage important public dialogue.

An ethical quandary is created when copyright enforcement activities infringe on fair use rights. The goal of fair use is to allow for the moral use of copyrighted content for things like criticism, commentary, and education. However, overzealous enforcement actions may have a chilling effect on fair use, deterring people and organizations from participating in actions they perceive to be morally acceptable. This conflict highlights the need for prudent ethical judgment in the enforcement field.

"Copyright trolling" is a problem that adds another ethical complexity. Copyright trolls abuse the legal system by sending threatening letters or requesting payments from those charged with copyright infringement, sometimes without substantial proof. This business-driven approach has come under fire for the possibility of misuse, which might place excessive financial and emotional strain on individuals unintentionally trapped in its web. Aiming for equity, proportionality, and the maintenance of moral limits in the digital environment, ethical copyright enforcement requires taking into account the interests of rights holders as well as the larger public good.

6.3 ETHICAL IMPLICATIONS OF PLAGIARISM AND CONTENT THEFT

Plagiarism and content theft present more intricate and pervasive ethical problems than ever before. Plagiarism is a grave violation of academic, creative, and professional ethics. It is widely described as utilizing someone else's work, ideas, or IP without due acknowledgment or permission. Contrarily, content theft is the illicit duplication or dissemination of digital material, including books, pictures, music, and software. Understanding the repercussions of plagiarism and content theft is essential in today's connected world since both ethical issues are quite serious.

The act of plagiarizing, a word derived from the Latin "plagiarius," which means "kidnapper" or "plunderer," strikes at the very foundation of intellectual honesty and integrity. It includes using someone else's creations, ideas, or IP without giving due credit or obtaining the required authorizations. Direct copying, paraphrasing without credit, self-plagiarism, patchwriting, and insufficient citation are just a few ways this unethical activity appears (Table 6.1) [3–5].

TABLE 6.1
Forms of unethical behavior related to plagiarism

Form of Unethical Behavior	Description
Direct Copying	Replicating someone else's text, code, artwork, or ideas verbatim without citation.
Paraphrasing Without Attribution	Rewrite someone else's work in your own words, but need to acknowledge the source.
Self-Plagiarism	Recycling your work, such as submitting the same paper or article to multiple outlets without disclosure.
Patchwriting	Combining fragments of various sources into a new document while maintaining the original wording.
Inadequate Citation	You need to properly credit the source even if you include it in your work.

Plagiarism has serious repercussions that extend well beyond academic and professional borders. Academic institutions implement severe sanctions to protect academic integrity, ranging from failing assignments to expulsion. Plagiarism may damage one's image both personally and professionally in a professional context, and once trust has been broken, it cannot be easy to regain. Plagiarism may have legal ramifications, especially when it contains copyrighted content. It prevents people from developing their ideas, which stifles creativity and innovation. It undermines confidence in various industries, including media and academic research, and confronts people with moral conundrums as they struggle with realizing their dishonest actions.

6.3.1 CONTENT THEFT IN THE DIGITAL ERA

Unprecedented potential for content production, dissemination, and consumption has been made possible by the digital age, but it has also presented a serious problem: content theft. Information theft, often known as digital piracy or copyright infringement, refers to various illegal actions, including duplicating, disseminating, and using digital information without the owner's permission. This problem has significant ethical, legal, and economic ramifications in today's linked society.

In the digital age, content theft may take many different forms, including unlawful streaming, file sharing, the creation of counterfeit products, and online content plagiarism. By denying producers fair pay and violating their IP rights, these actions reduce their motivation to develop high-quality work. In addition, copyright theft causes publishers, artists, and content providers to lose money, which might endanger their livelihoods and discourage innovation and creativity. It fosters an unjustified competitive environment where those participating in piracy have an edge over those who uphold ethical norms and copyright regulations.

Beyond moral issues, content theft has legal repercussions, such as copyright infringement lawsuits, fines, and the potential for criminal prosecution for people or organizations engaged in the unlawful distribution or use of intellectual works. DRM solutions, which seek to shield digital material from unlawful copying and distribution, are used by content producers and distributors to prevent content theft. DRM, however, also prompts moral issues about privacy and consumer rights. In order to combat copyright theft in the digital age, a careful balance must be struck between upholding authors' rights, encouraging innovation, and ensuring that moral behavior predominates online. It emphasizes protecting IP and encouraging responsible content use in a connected society.

6.3.2 Academic Integrity and Plagiarism

Academic honesty, fairness, and ethical information usage are the pillars around which all educational institutions are built. Plagiarism becomes a clear ethical concern when people portray someone else's work, ideas, or intellectual IP as their own in academic contexts. This dishonest behavior takes many forms, such as verbatim copying and inappropriate citation, and all compromise the fundamental principles of academic honesty.

Academic plagiarism has serious repercussions that go well beyond a single project. Universities and colleges react to plagiarism with severe sanctions, such as failing a course, failing an assignment, or even being expelled. Furthermore, it devalues academic degrees obtained dishonestly and degrades the educational experience by prohibiting students from acquiring critical skills. People who plagiarize often wrestle with internal moral conundrums due to the loss of confidence among students, teachers, and institutions and the realization that their integrity has been compromised.

Educational organizations and instructors are essential in the fight against academic plagiarism. Clear academic integrity rules are established; teaching materials are provided to assist students in understanding and avoiding plagiarism; techniques are used to identify plagiarism settings that encourage creativity; and ethical research methods are fostered.

6.3.3 Strategies for Combating Plagiarism Ethically

Education organizations, teachers, and students all share responsibilities for efficiently combating plagiarism while keeping moral standards. It entails a multifaceted strategy emphasizing prevention, instruction, and developing a culture promoting academic integrity.

The importance of education and awareness campaigns cannot be overstated. Faculty training enables educators to spot and resolve plagiarism while encouraging originality in projects, and plagiarism workshops and seminars may educate students about what plagiarism is and its repercussions. Policies on academic integrity must be clear. The definition of plagiarism and the consequences of infractions should be clearly stated in well-defined policies. These requirements must be stated up front in the course syllabus. Promoting ethical research and writing techniques is also essential. The temptation to plagiarize may be greatly reduced by teaching students the right citation formats and fostering critical thinking and analysis in assignments.

Such software may spot possible plagiarism, but it should be used mainly to teach students how to cite sources correctly, with thorough comments on the results.

Furthermore, it is crucial to promote an ethical culture. Maintaining a fair academic environment may be easier by developing an institutional culture prioritizing ethics and academic integrity, rewarding students who preserve these values, and promoting open communication about plagiarism.

Plagiarism may be decreased by creating genuine examinations that demand creative thinking and knowledge application to practical situations. Students should have easy access to counseling services and academic assistance resources like tutoring and writing centers. Finally, it is important to provide an example of moral conduct. Educators and administrators should emphasize the value of integrity via their academic work and relationships with pupils. The efficacy of plagiarism protection techniques is maintained by routinely evaluating and revising them in light of new developments and user input.

6.4   OPEN-SOURCE SOFTWARE AND ETHICAL COLLABORATION

A paradigm change in software development has been brought about by open-source software. Open-source software is based on openness, accessibility, and cooperation in contrast to proprietary software, which is held by a single corporation and often closed to outside contributions. Open-source software is any program made publicly accessible and whose source code is open for anyone to read, alter, and distribute. This openness creates a distinctive environment where moral cooperation is prioritized.

6.4.1   ETHICAL FOUNDATIONS OF OPEN SOURCE

Open-source software is transforming the software development environment by serving as both a development paradigm and a philosophy firmly grounded in ethical ideals. Open source is fundamentally committed to the principles of inclusivity and freedom. It promotes inclusiveness by allowing people to read, change, and share source code. It advocates for open-source software. The belief that technology should not be a privilege but a shared resource is strengthened by democratizing technology, which gives people and communities more influence.

Community and cooperation are key components of open source. Developers from all backgrounds are united, fostering trust, support, and shared accountability. Fundamental concepts are accountability and transparency, with open source highlighting the need to enable anybody to see the source code. Making software developers accountable for the quality and security of their work promotes ethical conduct in the software development industry. The usage and distribution of open-source software are strictly regulated by ethical licensing frameworks like the GNU General Public License (GPL), ensuring that transparency and cooperation are respected.

Open source extends its concept to information sharing in addition to coding. Developers make their knowledge and experiences available to the larger community along with their code. This culture of information exchange encourages education, creativity, and moral progress in technology. Furthermore, open source may result in more moral and ethical business practices by lowering software-related expenses, fostering competition, and minimizing resource consumption and waste.

6.4.2   Collaborative Development and Ethical Considerations

Open-source software is characterized by its collaborative development methodology, which cooperatively brings together people and communities to produce and improve software. This collaborative strategy is fundamentally moral, reflecting the values of diversity, reciprocity, and ethical technology management. In open source, diversity flourishes as contributors from many origins, ethnicities, and skill levels are warmly welcomed, encouraging creativity by embracing a broad range of viewpoints. In contrast to more restricted, private development environments, this inclusion, grounded in ethics, lowers obstacles and promotes equitable possibilities for participation.

The open-source community strongly emphasizes ethical principles that respect individual contributions. Each contributor's work is respected and acknowledged, protecting their IP rights and fostering trust. Additionally, community norms and standards of conduct serve as crucial moral compass points by establishing expectations for civil behavior, inclusive communication, and efficient dispute resolution. Another fundamental principle of ethical open-source development is transparency in decision-making. This principle guarantees that significant choices are taken jointly and with public involvement, limiting undue influence and covert agendas.

The two most important ethical issues in collaborative open source are security and responsible disclosure. The security and integrity of software are given top priority in projects, and the appropriate disclosure of security flaws is stressed. Ethical licensing decisions preserve the program's independence and openness in keeping with the fundamental tenets of open-source software. Open-source initiatives attempt to prevent exploitation while promoting fruitful cooperation driven by clear principles and ethical limits and balancing interactions with corporate organizations and economic interests. Lastly, long-term upkeep and sustainability planning are crucial ethical considerations that guarantee projects will continue to be functional, accessible, and advantageous to future generations.

6.4.3   Benefits and Challenges of Open Source

Although open-source software is praised for its many advantages, it also has several disadvantages that businesses and people should be aware of. The main benefit of open source is its affordability since it does away with the necessity for pricey software licensing. Startups, educational institutions, and charitable organizations are drawn to this affordability because it enables them to distribute resources more effectively. Additionally, open source provides flexibility and customization, enabling users to modify software to meet their unique requirements and boosting workflow effectiveness.

A worldwide community of developers' combined efforts make open-source software more transparent and safe. Security breaches are less likely due to early vulnerability detection and remediation. Collaboration across open-source groups encourages creativity, which produces new features and quick software updates. Organizations have more control over their software solutions because of the lack of vendor lock-in, which encourages independence and flexibility.

Open source does, however, provide a unique set of difficulties. It might not be easy to locate suitable support and maintenance services, especially for vital corporate systems. Productivity may be impacted by learning curves and the requirement for employee training. When attempting to integrate open-source technologies into current IT infrastructures, compatibility and integration problems may occur. It might not be easy to ensure quality and that open-source projects adhere to strict requirements. Finally, it is essential to comprehend open-source licenses and their ramifications to prevent legal entanglements and IP issues.

Finally, there is no denying that open source offers cost reductions, flexibility, security, and creativity advantages. To fully use the benefits, however, enterprises must manage support, compatibility, quality control, and licensing difficulties. For open source to be successfully adopted and integrated into various settings, a thorough grasp of its advantages and disadvantages, as well as careful planning and continuous management, are necessary.

6.5   BALANCING IP RIGHTS WITH INNOVATION

IP rights and the need for innovation are increasingly at odds. The protection of individual and corporate inventions depends heavily on IP, which includes copyrights, patents, trademarks, and trade secrets. Striking a balance between upholding these rights and promoting innovation for the greater good is a difficult ethical challenge raised by the quick speed of technological development and the constantly growing scope of digital information.

6.5.1   THE INNOVATION-IP RIGHTS CONFLICT

The conflict between IP rights and innovation has become more apparent, and it is now known as the "Innovation-IP Rights Conflict." This issue highlights the difficult moral decisions that must be made to safeguard IP while still fostering innovation for the good of society.

A fundamental paradox lies at the core of the conflict between innovation and IP rights: while strong IP rights are necessary to encourage creators and innovators to devote their time, energy, and creativity to creating new concepts, technologies, and creative works, overly strict IP protection can impede development and stifle innovation. This dilemma illustrates the fine line that must be drawn to uphold both individual rights and the larger goals of innovation.

The fundamental distinction in the argument is between exclusive ownership and shared knowledge. IP laws provide inventors and artists exclusive ownership rights over their IP, allowing them to manage and monetize their works. This exclusivity acts as a draw, encouraging people and businesses to spend money on novel ideas.

The problem arises, however, when these exclusive rights are unduly limiting. This could hide concepts and information from the general public, limiting others from expanding on or inventing in related fields. It raises moral concerns regarding whether protecting the creator's rights or the benefit of society should be the main objective.

It is difficult to navigate the Innovation-IP Rights Conflict, which requires a constant focus on moral issues. The delicate balance requires constant attention, which calls for a careful analysis and adaption of governmental policies and commercial procedures. The need to regularly assess the extent of IP protection, ensuring that it complies with the dual goals of promoting innovation and protecting creators' rights, acknowledging the varied manifestations of the conflict across various domains, such as technology, entertainment, and academia, and advocating transparency within IP procedures are just a few of the important aspects of ethical decision-making in this area.

6.5.2 ETHICAL DILEMMAS IN PATENTING AND INNOVATION

Patents have raised moral issues about exclusivity, accessibility, and fair competition while being intended to protect inventors and promote innovation. Patents have a dual function: they provide inventors exclusive rights while requiring thorough disclosure, advancing general knowledge. However, unfair patent activities, such as patent trolling, which targets startups and smaller inventors by abusing patents for litigation and licensing fees, pose a danger to hinder innovation.

It is critical to continuously evaluate the breadth of IP protection to ensure it adheres to the goals of fostering innovation while defending authors' rights to manage these moral dilemmas. Promoting cooperation and transparency, advocating fair licensing, and combating monopolistic tactics that restrict accessibility are some strategies. Additionally, open innovation models, defensive patenting, and patent reform all provide ways to address moral quandaries in innovation and patenting.

6.5.3 PATENT TROLLING AND ITS ETHICAL IMPLICATIONS

The practice of patent trolling, also known as patent assertion entities (PAEs) or non-practicing enterprises (NPEs), has sparked serious ethical questions in the area of IP rights [6, 7]. It includes obtaining patents only for suing or collecting license fees rather than for invention or commercialization. This situation puts the purpose and effects of patent ownership under intense ethical scrutiny.

Patent trolling is fundamentally uncommitted to innovation. Instead, it employs legal strategies to make money off of the innovations of others, thereby impeding real progress by scaring inventors and restricting healthy competition. This approach places a heavy burden on startups and small firms since they often need more means to fight out drawn-out legal battles, discouraging entrepreneurship and hindering economic progress. Furthermore, the judicial system is clogged by the rise of frivolous litigation brought by patent trolls, raising questions about the responsible use of patents and their intended function in fostering advancement.

A diversified strategy is required to address the moral concerns of patent trolling. This entails lawmakers continually reviewing patent laws to stop and encourage age openness and ownership. The ethical licensing procedures patent owners use should balance upholding their legal rights and promoting innovation and competition.

6.5.4 PROMOTING INNOVATION WHILE RESPECTING IP RIGHTS

Striking a balance between preserving IP rights and promoting innovation is a fundamental challenge in the current digital environment. In this context, the ethical considerations coalesce around striking a delicate balance between encouraging creativity, technological advancement, and the dissemination of knowledge while protecting the rights of creators and innovators. IP rights, such as patents, copyrights, trademarks, and trade secrets, are crucial incentives for research and development investment, resulting in ground-breaking innovations and creative works. However, the intricate relationship between these rights and innovation necessitates ethical strategies to ensure that innovation flourishes within the framework of IP protection.

Selectively enhancing IP enforcement is a crucial strategy. While robust IP protections are essential, they should be enforced judiciously to prevent hindrance to creativity and competition. Targeting flagrant IP infringement and misuse while permitting more flexible interpretations in other instances is a prudent strategy. In addition, fostering practices such as fair use and open access promotes knowledge sharing and creative expression while protecting IP rights. Collaboration between IP holders and innovators, ethical licensing practices, and adopting open innovation models all contribute to establishing an ethical balance between IP protection and innovation promotion. Fostering innovation while respecting IP rights is, in essence, an ongoing ethical endeavor that requires adaptability, vigilance, and a commitment to the broader societal benefits of progress and knowledge diffusion.

6.6 ETHICAL PRACTICES IN LICENSING, PATENTS, AND TRADEMARKS

The ethical issues relating to licensing, patents, and trademarks are of utmost significance in the IP field. For the rights of artists to be protected, innovation to be encouraged, and fair market competition to be ensured, these facets of IP law are essential. They also need to balance upholding these rights and abstaining from immoral behavior carefully. This section examines the moral implications of trademarks, patents, and licensing.

6.6.1 LICENSING AND ETHICAL BUSINESS PRACTICES

Giving or receiving permission to utilize protected works of art, innovations, or trademarks is vital to managing IP. It is a potent instrument for leveraging intellectual assets for the gain of individuals, companies, and organizations. But the moral issues involved in licensing are just as crucial as the legal ones. Transparency and clarity in all contracts are the first steps toward ethical licensing. To make wise judgments, parties to a license agreement should be well aware of its terms and circumstances. This entails using plain language in contracts, eliminating legalese or jargon, and ensuring all parties can access attorneys if necessary.

Fairness and equality are ideals that ethical licensees and grantors pursue in their contracts. For the use of their IP, creators should be fairly compensated, and licensees should be given value in line with their investment. Unmoral actions, such as engaging in dishonest negotiations or charging high rates, may undermine trust and harm both parties.

Ethical licensing avoids imposing too onerous conditions that hinder ingenuity and creativity. Licensors are entitled to safeguard their IP, but they are also responsible for thinking about how their decisions may affect other people. Overly onerous licensing conditions may impede development, restrict competition, and impede the spread of information. The best licensees strike a balance between preservation and development.

Collaboration between IP developers and users is a common feature of licensing agreements. The collaborative spirit is acknowledged and respected in ethical practices. This entails creating an atmosphere where both parties may gain from one another, learn from one another, and aid in developing their sectors. Ethical licensors support this dynamic since it may result in innovation and the creation of fresh solutions.

Exploitative conduct, in which one party takes advantage of the other's weaknesses or ignorance, is prevented by ethical licensing. This might include exploitative licensing methods, hidden clauses, or unjust pricing. Licensors and licensees that act ethically participate in talks with honesty and fairness, ensuring that all parties are aware of the agreement's ramifications. Table 6.2 summarizes the moral issues, fundamental ideas, significance, and illustrations relating to major facets of corporate conduct and licensing in IP agreements.

TABLE 6.2
Key considerations in ethical licensing and business practices for IP

Aspect of Ethical Licensing	Key Considerations	Importance	Examples
Transparency and Clarity	– Understandable language in agreements. – Access to legal counsel for all parties.	High	– Providing a plain-language summary of terms alongside legal jargon.
Fairness and Equity	– Fair compensation for creators.	High	– Ensuring revenue sharing in a licensing deal is equitable.
Avoiding Overly Restrictive Terms	– Balancing IP protection with innovation. – Avoiding stifling effects on creativity.	Moderate	– Allowing for flexibility in usage within the licensing agreement.
Respecting the Spirit of Collaboration	– Encouraging cooperation and knowledge sharing. – Promoting a win-win approach in agreements.	High	– Encouraging knowledge sharing and joint development in agreements.
Preventing Exploitative Behavior	– Avoiding harm to other parties in agreements. – Ensuring honest and fair dealings.	High	– Avoiding monopolistic control through aggressive licensing.

6.6.2   ETHICAL CONSIDERATIONS IN PATENT ACQUISITION

Maintaining the integrity of the patent system depends heavily on ethical issues in the purchase of patents. The difference between real invention and patent trolling is at the core of this problem. In order to get a patent ethically, one must be dedicated to supporting genuine scientific improvements rather than pursuing financial gain via litigation. Full disclosure and transparency are essential; inventors must provide accurate descriptions of their innovations to maintain the credibility of patent records for the public and other inventors. Additionally, it is ethically required to perform exhaustive research to find "prior art" in order to stop the issue of patents for well-known technology.

Another aspect of ethics is balancing rights and the general good. Patents provide creators the right to exclusivity, but they also should progress technology and society. Instead of hoarding or exploiting their patents just for vindictive litigation strategies, ethical patent owners actively support innovation and teamwork. Lastly, moral patent acquisition requires avoiding unethical tactics like submitting frivolous or too wide patent applications that restrict innovation and competition. The key to ethical patent acquisition is responsible and careful patent filing, which ensures that the patent system continues to promote genuine innovation as intended.

6.6.3   TRADEMARKS AND ETHICAL BRANDING

In business, trademarks, which include distinctive symbols, names, or designs, act as potent identifiers. They build customer trust and brand awareness, making ethical branding an important factor. At its foundation, ethical branding strongly emphasizes honesty and accuracy, ensuring that trademarks accurately reflect the goods and services they are linked to. In addition to damaging customer trust, misleading or deceptive branding runs the danger of legal ramifications.

Ethical branding calls for consideration of social and cultural standards. It obliges businesses to abstain from disrespectfully or offensively using cultural or religious symbols and to prevent unintentionally upsetting certain communities with their trademarks. Another important consideration is using trademarks responsibly, which includes avoiding linking them to goods or services that go against societal norms or are associated with unfavorable or contentious causes.

Enforcement of trademarks is essential to ethical branding. Although trademark protection is crucial, it must be done equitably and proportionately. Abusing trademark rights to quiet rivalry or criticism may have bad public impression effects and legal repercussions. Additionally, ethical branding includes sustainability and environmental responsibility when raising environmental awareness. Companies should refrain from deceiving customers with exaggerated claims about sustainability and honestly express their dedication to eco-friendly activities in their trademarks.

Ethical branding includes civic engagement and charity. Companies should be open and sincere when associating their brands with humanitarian endeavors or social issues. Without making significant donations, using trademarks to promote charity might be exploitative and immoral.

6.6.4 NAVIGATING IP DISPUTES ETHICALLY

IP conflicts are a natural component of the environment for creativity and innovation. Ethical concerns play a crucial role in settling disputes that involve the ownership, infringement, or abuse of IP. Maintaining standards of justice, openness, and respect for the rights of all parties concerned is necessary for ethically resolving IP conflicts.

Using Alternative Dispute Resolution (ADR) techniques like negotiation, mediation, or arbitration is one moral way to settle IP conflicts [8]. ADR emphasizes a commitment to resolving conflicts constructively and ethically by allowing parties to look for peaceful solutions outside of expensive and drawn-out litigation. However, ethical behavior continues to be crucial when litigation is inevitable. Lawyers and parties must uphold the highest ethical standards throughout the legal process, provide accurate information, and refrain from abusive litigation strategies.

A fine balancing act must be struck between upholding one's rights and recognizing fair usage or the public interest to navigate IP challenges ethically. Practitioners must consider how their decisions may affect the public domain, creativity, and innovation in general. Parties are required to communicate pertinent information and supporting documentation in order to preserve ethical integrity. The pursuit of fair and just settlements above strategies intended to stall or obstruct the process is highlighted by a commitment to resolving disputes in good faith, which emphasizes ethical behaviors in IP disputes.

REFERENCES

1. Kapitsaki, G.M. and G. Charalambous, Modeling and recommending open source licenses with findOSSLicense. *IEEE Transactions on Software Engineering*, 2019. 47(5): pp. 919–935.
2. Stokes, S., *Digital copyright: Law and practice*. 2019: Bloomsbury Publishing.
3. Thurmond, B.H., *Student plagiarism and the use of a plagiarism detection tool by community college faculty*. 2010: Indiana State University.
4. Vrbanec, T. and A. Meštrović, Taxonomy of academic plagiarism methods. *Zbornik Veleučilišta u Rijeci*, 2021. 9(1): pp. 283–300.
5. Sharma, H. and S. Verma, Insight into modern-day plagiarism: The science of pseudo research. *Tzu Chi Medical Journal*, 2020. 32(3): pp. 240–244.
6. Allison, J.R., M.A. Lemley, and D.L. Schwartz, How often do non-practicing entities win patent suits? *Berkeley Technology Law Journal*, 2017. 32(1): pp. 237–310.
7. Valentine, Z.H., A novel, nonobvious approach to curb abusive patent litigants. *Roger Williams University Law Review*, 2016. 21: p. 118.
8. Menkel-Meadow, Carrie J., Mediation, Arbitration, and Alternative Dispute Resolution (ADR) (May 19, 2015). *International Encyclopedia of the Social and Behavioral Sciences, Elsevier Ltd*. 2015, UC Irvine School of Law Research Paper No. 2015-59, Available at SSRN: https://ssrn.com/abstract=2608140

7 Ethical Software Development and Testing

7.1 ETHICAL CONSIDERATIONS IN SOFTWARE DEVELOPMENT LIFECYCLE

Ethical deliberations constitute a fundamental component of academic writing and research, guaranteeing that investigations are carried out with accountability and integrity. Researchers must uphold ethical considerations during their investigations, including confidentiality, potential for injury, informed consent, anonymity, and conflicts of interest. Researchers must conduct their investigations in adherence to the optimal methodologies and protocols established by diverse international organizations and authoritative bodies. Good Clinical Practice (GCP) is a globally recognized standard of quality that establishes directives for implementing clinical research involving human subjects [1].

When formulating the research methodology, scholars must incorporate ethical considerations, which serve as the foundation for the credibility of scholarly discussions. In order to advance research objectives, safeguard scientific integrity, and encourage public participation in policy-making, ethical considerations are indispensable. Noncompliance with ethical considerations could result in public backlash and impede endeavors to establish policies about research findings.

Researchers must guarantee that their investigations are carried out ethically and accountably, from planning to disseminating findings [5]. Participants' informed consent is a critical component of ethical research; they must possess comprehensive knowledge regarding the nature of the inquiries, the intended use of the data, and the potential repercussions. Additionally, researchers are responsible for safeguarding the identities and responses of research participants. In health research, for instance, confidentiality is crucial because a violation of confidence could stigmatize participants afflicted with a malady.

The significance of ethical considerations lies in ensuring the validity of research, protecting participants' rights, and fostering an environment of collaboration, trust, and mutual respect between participants and researchers. Research findings are rendered invalid, and participants endure severe physical, social, and psychological injury when unethical practices are implemented. Researchers must adhere to ethical considerations when gathering data and proposing findings that can inform policy-making.

7.1.1 INTEGRATING ETHICS INTO REQUIREMENTS GATHERING

Integrating ethical considerations into the requirements-gathering process is essential to the software development lifecycle. Numerous scholars have investigated the

 DOI: 10.1201/9781003584452-7

ethical implications of software development processes, software engineering codes of ethics, and professional practices [2]. When integrating ethics into requirements gathering, it is critical to consider the techniques and methods utilized to identify ethical issues throughout the process. This may encompass the formulation of a research protocol that undergoes thorough consensus among all participants and associates, alongside the implementation of data collection instruments that explicitly acknowledge ethical considerations. In addition, it is imperative to secure ethics approval before collecting data from human participants; doing so may result in unethical behavior. As a result of societal expectations for greater accountability, there has been a heightened focus on ethical behavior, underscoring the significance of incorporating and overseeing ethical concerns at every stage of the research endeavor.

Comprehending and implementing the fundamental ethical principles of autonomy, justice, beneficence, and nonmaleficence in software engineering requirements gathering is crucial. Emphasis should be placed on autonomy, the foundation for truth-telling, informed consent, and confidentiality, when gathering requirements from stakeholders [3]. In addition, personal and professional conflicts of interest ought to be managed and reported transparently to safeguard the integrity of the requirements-gathering procedure. This is particularly significant when researchers conduct research related to their institutional programs or have prior relationships or activities that could give rise to a conflict of interest.

Consent with knowledge is a fundamental principle of ethical investigation and should be meticulously weighed when soliciting information from human subjects. The informed consent procedure guarantees that participants are completely enlightened regarding the objectives and potential ramifications of the research and that they have explicit consent to partake. This includes ensuring they are cognizant of their entitlements to access their information and disengage from the study at any moment. This is especially crucial in software requirements engineering, where the input and approval of stakeholders are critical to the project's success.

7.1.2 ETHICAL DECISION-MAKING IN DESIGN PHASE

Ethical decision-making is important during the design phase across various disciplines, such as management, behavior analysis, and dentistry. Over time, numerous ethical decision-making models have been formulated and implemented, such as the systems-oriented ethical decision-making framework, the Four-Box Method, and the narrative dental ethics decision-making model. The models above comprise many stakeholder concerns and ethical principles that challenge contemplation. Their objective is to provide a methodical ethical decision-making framework [4–6].

Several stages comprise the ethical decision-making process: ethical radar, problem identification, information gathering, evaluation of available options/behaviors, and decision-making. The intricate procedure entails variable interactions that may be evaluated in the field. Enhancing professionals' comprehension of the process can facilitate formulating policies that increase the probability of ethical conduct within their respective organizations [7]. Implementing a systems-oriented ethical decision-making framework can greatly assist in addressing the negative impacts of structural and social determinants of health. By incorporating deontological and utilitarian ethical principles, the framework facilitates the growth of empathy. This resource assists health system leaders, organizations, and providers in navigating the progressively intricate ethical aspects of providing care for underserved patients disproportionately affected by social and structural determinants of health [4].

7.1.3   Code Development with Ethical Principles

Professionals adhere to ethical codes and conduct to ensure that their endeavors are consistent with moral and societal values. Software development, an industry where the influence of technology on society and individuals is substantial, is where these principles are especially crucial. In order to ensure ethical software development, a code of ethics must be incorporated throughout the software development life cycle. A comprehensive understanding of the ethical considerations linked to software engineering procedures and the techniques employed to guarantee the preservation of significant ethical issues during the development phase is necessary. Software engineers can ensure that their decisions prioritize the public interest and consider the potential repercussions of their work on both individuals and society by abiding by ethical codes [2].

Professionals in the field of software requirements engineering encounter a multitude of ethical and moral quandaries. A meticulous examination of privacy, security, and the potential societal ramifications of the software under development is necessary to incorporate ethical considerations during this stage of development [2]. Software engineers can promote the development of technology ethically and responsibly by explicitly addressing these concerns throughout the requirements of the engineering process.

7.1.4   Ethical Considerations in Documentation and Communication

Ethical considerations are paramount in promoting accountability, transparency, and responsible development practices. Ethical documentation requires the preservation of information integrity, whereby assertions regarding the capabilities and limitations of the system are faithfully reflected. The promotion of inclusive language is prioritized, with an effort to avoid perpetuating biases and stereotypes in order to accommodate diverse audiences with differing levels of technical expertise.

Transparency is a fundamental ethical principle that significantly emphasizes open and honest communication. It is imperative that developers candidly disclose any identified concerns, hazards, or constraints in the software, thereby fostering user confidence and facilitating well-informed choices. Additionally, user privacy and data protection should be prioritized, with documentation that explains the software's purpose, security measures, and how it handles user data. Implementing comprehensive security guidelines enhances user safety by providing explicit directives about setting configuration, access control management, and vulnerability resolution.

Additionally, documentation should establish channels of ethical communication for user concerns and feedback. Clear instructions regarding reporting issues, providing feedback, or seeking assistance are essential for users. Prompt and courteous replies indicate a dedication to ethical participation. In brief, ethical documentation and communication entail the observance of principles such as upholding integrity, employing inclusive language, encouraging transparency, placing user privacy as a top priority, furnishing security guidelines, and cultivating responsive communication channels. By adhering to these principles, software development can maintain its ethical foundation and remain in line with core values such as user-centricity, transparency, and integrity (Figure 7.1).

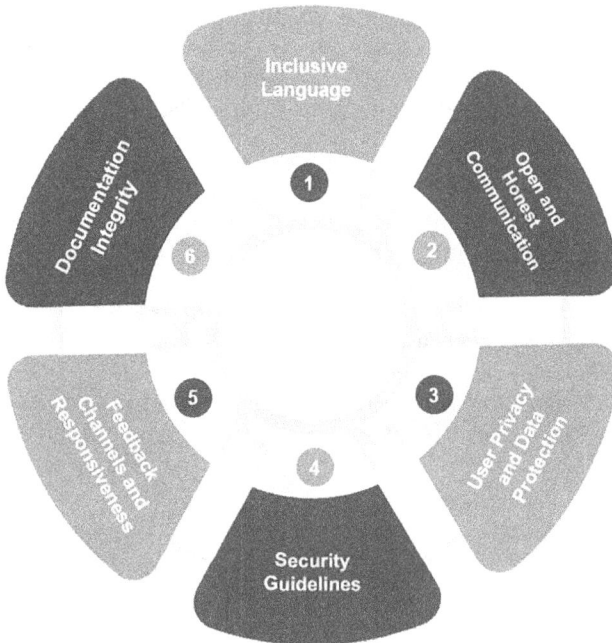

FIGURE 7.1 Core values for ethical practices

7.1.5 MONITORING AND ENFORCEMENT OF ETHICAL STANDARDS

The enforcement of ethical standards necessitates the implementation of resilient mechanisms that monitor and evaluate the ethical ramifications of the software throughout its lifecycle. This includes evaluating the impact on end-users conducting code evaluations and ongoing monitoring of development processes. Implementing routine audits can effectively detect and promptly rectify any instances of ethical standards being violated. Through the implementation of monitoring systems, development teams can proactively track ethical performance and make necessary adjustments to ensure adherence to ethical principles.

In order to enforce ethical standards, policies and procedures addressing ethical violations must be established. This may entail the establishment of an ethics committee with the specific responsibility of assessing and addressing reported ethical concerns. Consequences for ethical transgressions should be detailed, emphasizing the criticality of ethical behavior during the software development lifecycle.

The enforcement procedure should be open and transparent to facilitate a culture of accountability for the development team. Consistent education regarding ethical principles and the repercussions for failing to comply can strengthen the significance of ethical conduct and cultivate an environment that places ethical deliberation at the forefront of decision-making processes.

In addition to monitoring and enforcement initiatives, organizations may incorporate ethical impact assessments into the software development lifecycle. Systematically evaluating the potential ethical ramifications of a software project prior to, during, and after its development constitutes these assessments. Adopting a proactive approach facilitates the early identification and resolution of ethical concerns, thereby reducing the likelihood of inadvertent ethical infringements and bolstering the software product's overall ethical integrity.

7.2 BALANCING USER NEEDS, BUSINESS INTERESTS, AND ETHICAL RESPONSIBILITIES

Contemporary organizations encounter a formidable obstacle in reconciling the intricate dynamics that span user requirements, business objectives, and ethical obligations. Scholarly investigations indicate that business models frequently entail conflicts between satisfying customer demands and attaining financial prosperity [8]. Furthermore, ethical responsibility incorporates many facets, such as the preservation of the environment, the implementation of equitable labor practices, and the observance of individual rights [9].

Adhering to ethical standards in user experience (UX) design entails guaranteeing that products satisfy users' requirements and accomplish organizational objectives while upholding their integrity. UX designers must balance business objectives and user experience, maintaining user confidence and contentment. There is a growing trend among organizations to integrate ethics into their fundamental operations, shifting their attention from economic gain to more extensive societal issues. Organizations such as the Business Roundtable have expanded their objectives to encompass the welfare of all stakeholders instead of focusing solely on shareholders' interests [9].

Entrepreneurs encounter distinct obstacles when reconciling ethical deliberations with pursuing progress and expansion. Acknowledging the intricacy of ethical entrepreneurial decisions and conduct is crucial to promoting sustainable development [10].

7.2.1 IDENTIFYING USER NEEDS AND EXPECTATIONS

User requirements recognition necessitates an all-encompassing comprehension of the end-users, their varied origins, and the environment in which they will engage with the software. In order to obtain a comprehensive understanding of the intended user base's needs, preferences, and obstacles, developers must undertake extensive user research, including surveys and interviews. By doing so, developers can cultivate a more user-centric and ethical approach to software design by ensuring their work aligns with users' practical requirements.

Furthermore, identifying user expectations encompasses ethical considerations, user experience, and functional requirements. There is a growing trend among users to require software that satisfies their pragmatic requirements and adheres to their ethical and value systems. Developers must be cognizant of these expectations and incorporate security, inclusivity, and data privacy into their design methodology. Ethical considerations, including a dedication to accessibility and transparency regarding user data handling, are crucial to fulfilling and surpassing user expectations. In order to develop software that satisfies the ethical expectations of its diverse user base and technically meets specifications, achieving a harmonious equilibrium among functionality, user experience, and ethical principles is critical.

Persisting communication and feedback mechanisms are critical to ascertain user requirements and expectations. Consistently interacting with users, soliciting their feedback, and modifying software accordingly constitutes a dynamic procedure that guarantees the ongoing congruence between software development and the ever-changing demands of users. Developers can maintain their receptiveness to evolving user demands, technological progress, and emergent ethical dilemmas by employing this iterative methodology.

7.2.2   ALIGNING BUSINESS INTERESTS WITH ETHICAL GUIDELINES

Within the intricate realm of software development, organizations are compelled to navigate a precarious equilibrium between their financial objectives and ethical deliberations. This entails establishing and adopting ethical business practices that place a premium on integrity, transparency, and accountable behavior. Incorporating ethical principles into the formulation of strategic plans guarantees the initial consideration of ethical ramifications, harmonizing corporate objectives with societal values. Furthermore, the involvement and cooperation of stakeholders are essential in the early detection of ethical hazards, which enables the implementation of preventative actions to resolve and alleviate apprehensions. Organizations can cultivate a workforce attentive to and dedicated to ethical principles by allocating resources toward ethics training programs and integrating ethical metrics into performance evaluations. Ongoing assessment and adjustment of business practices are imperative to guarantee conformity with progressive ethical benchmarks, bolster stakeholder confidence, and positively contribute to an ethically responsible and sustainable software development ecosystem.

7.2.3   NAVIGATING CONFLICTS BETWEEN STAKEHOLDERS

Disagreements over project direction, feature priority, and resource allocation can arise from stakeholders' varied viewpoints, goals, and expectations. Project managers and development teams need to resolve problems in this environment by working together and being open and honest. A transparent decision-making process, frequent meetings, and open lines of communication are necessary to guarantee that all parties' concerns are considered.

Establishing an ethical decision-making framework that considers all pertinent parties' opinions is one practical method for handling conflicts. This might entail setting up a specific ethics committee or asking ethics experts outside the organization for unbiased advice. Software development teams may create agreement and avoid ethical problems by including stakeholders in decision-making. Documenting the decision-making process and the reasoning behind important decisions also improves accountability and transparency, builds stakeholder confidence, and lowers the likelihood of disputes. Giving ethical concerns precedence over immediate financial rewards when disputes continue is imperative. This might entail going over the project's objectives again, reevaluating its priorities, and, if needed, modifying them to conform to moral principles.

7.2.4 TRANSPARENCY AND OPEN COMMUNICATION IN DECISION-MAKING

Giving stakeholders easily available and understandable information on the decision-making process is a key component of transparent processes. This entails explaining the reasoning behind choices while considering user requirements and corporate objectives. Open and honest processes build trust among all parties involved, building strong bonds with customers, internal team members, and end users.

Open communication lines allow stakeholders to voice concerns, offer input, and raise ethical issues, which complements transparency. Regular meetings, feedback sessions, and open documentation can help achieve this. Promoting open communication allows for spotting possible ethical issues early in the decision-making process. It gives team members the confidence to participate in moral conversations without worrying about facing consequences.

Transparency in decision criteria is essential for giving each decision in the decision-making process a clear foundation. This entails specifying the standards by which features are ranked, technologies are chosen, and strategic choices are made. Teams build accountability and strengthen the legitimacy of their selected course of action by being transparent about their decision reasons. Furthermore, recognizing possible ethical ramifications and being transparent about trade-offs empowers stakeholders to evaluate the ethical issues on their own, promoting a culture of ethical decision-making and ongoing development based on stakeholder input.

7.3   QUALITY ASSURANCE AND ETHICAL TESTING PRACTICES

7.3.1 IMPORTANCE OF ETHICAL TESTING

This part of testing aims to find and address any ethical issues that may surface during the software development lifecycle, going beyond conventional functionality and performance tests. The importance of ethical testing is found in its capacity to solve privacy proactively, security, and fairness-related concerns, hence promoting an accountable and reliable software environment.

The growing use of technology in many facets of daily life is one of the main arguments in favor of placing a strong emphasis on ethical testing. Software programs gather and handle enormous volumes of personal data; therefore, protecting user privacy requires careful consideration of ethical issues. Ethical testing ensures that developers put strong security measures in place to shield customers from unwanted access and data breaches by helping to uncover flaws that might jeopardize sensitive information. It also helps reveal biases or discriminatory algorithm behavior, promoting equity and inclusion in software systems that may affect various user groups.

Moreover, the standing and reliability of software developers and companies are also impacted by ethical testing. Mistakes or unethical behavior can have serious repercussions, including legal repercussions and harm to a company's reputation. Organizations may foster confidence with users and stakeholders by prioritizing ethical testing to showcase their dedication to responsible and transparent software development. In an increasingly morally aware digital environment, this strategy protects against legal and financial problems and supports the software products' long-term viability.

7.3.2 INTEGRATING ETHICAL CONSIDERATIONS IN TEST PLANNING

The first stage in creating complete ethical test goals is to uncover algorithmic decision-making biases and concentrate on privacy and data security. This proactive approach guarantees that the software satisfies functional, moral, and societal requirements by assisting in the early discovery and resolution of ethical concerns.

Test planning should include a range of scenarios to mimic real-world settings and consider a wide user base (Figure 7.2). This inclusion goes beyond socioeconomic status, cultural background, and demography to confirm that the software acts morally toward various user groups. The testing method must adhere to strict criteria to handle sensitive data with care and respect for privacy to ensure the data's ethical usage. Two essential elements of ethical data usage are gaining user consent for testing purposes and adhering to data protection standards.

Making ethical decisions when doing tests is essential. Testing teams should have the skills and information necessary to put users' needs first and deal with moral issues as soon as they arise. Transparency is further improved by procedures for reporting and documenting ethical discoveries, which notify stakeholders promptly so that corrective action may be taken.

FIGURE 7.2 Ethical considerations in test planning flowchart

Ethical test cases entail assessing how the program behaves in situations in order to prevent and resolve. This procedure aims to identify any biases, discriminatory results, or unforeseen effects that could occur from using it in the actual world. For ethical test cases, Table 7.1 includes important factors, testing strategies, test scenarios, and tools emphasizing user accessibility, diversity, effect assessment, data management, informed permission, fairness metrics, unintended consequences, and continuous monitoring.

TABLE 7.1

Key considerations for conducting ethical test cases

Key Considerations	Testing Approach	Test Scenarios	Examples of Tools/Techniques
Diversity in Test Data	Create diverse datasets.	Test bias across user groups.	Data augmentation techniques.
	Assess system response to diverse inputs.	Identify disparities in outcomes.	Statistical analysis of data distribution.
Sensitive Data Handling	Test encryption for data at rest and in transit.	Simulate secure data transmission scenarios.	Penetration testing for security.
	Validate secure storage and retrieval.	Check for vulnerabilities in data handling.	Security audit of data handling processes.
Impact Assessment	Conduct impact analysis on diverse segments.	Evaluate effects on stakeholders and marginalized groups.	Stakeholder interviews and surveys.
	Implement tests for societal impact.	Address concerns related to social and economic impact.	Ethical impact assessments.
Informed Consent Testing	Assess the software's ability to manage consent.	Validate compliance with data protection regulations.	Usability testing of consent interfaces.
	Test consent dialogs and forms.	Ensure transparent and understandable consent mechanisms.	Legal review of consent processes.
User Accessibility	Use accessibility testing tools and guidelines.	Evaluate accessibility for users with various impairments.	Assistive technology compatibility testing.
	Conduct usability testing.	Address navigation, readability, and usability issues.	Accessibility audits of user interfaces.
Fairness Metrics	Establish criteria for fairness measurement.	Evaluate outcomes against predefined fairness metrics.	Statistical fairness measures.
	Implement tests for biased decisions.	Address disparities through algorithm adjustments.	Bias detection algorithms.
Unintended Consequences	Implement tests for unforeseen behavior.	Address issues arising from unforeseen interactions.	Chaos engineering for system resilience.
	Conduct stress and edge case testing.	Regularly update monitoring based on emerging trends.	Log analysis for unexpected patterns.
Continuous Monitoring	Implement automated ethical monitoring tools.	Establish feedback loops for rapid response.	AI-powered ethical monitoring systems.
	Regularly update monitoring based on trends.	Ensure consistent response to emerging ethical concerns.	Continuous integration of ethical checks.

7.3.3 Reporting and Addressing Ethical Issues Found in Testing

It is essential to have a strong reporting system in place in case ethical issues are discovered during the testing phase. This entails setting up a transparent and easily accessible avenue for testers to report any ethical problems they encounter while testing. It is advisable to motivate testers to furnish comprehensive records of the ethical issues, including the circumstances behind their emergence, consequences, and any corroborating data. This knowledge is essential for developers and stakeholders to comprehend the nature of the problem and take necessary action.

A methodical and open procedure for handling ethical concerns must be implemented as soon as they are reported. This entails conducting a comprehensive inquiry to confirm the veracity and seriousness of the concerns raised. The development, testing, and ethical teams must work together to comprehend the full implications of issues that have been found. To avoid possible harm, mitigating measures need to be put in place right once, depending on how serious the ethical problem is. This might entail stopping the software's distribution, putting in place interim solutions, informing users of the problems found, and offering advice on reducing any risks. Keeping the confidence of users and stakeholders during this process requires open and honest communication.

In addition, addressing ethical concerns in testing entails taking proactive action to avoid future occurrences of the same problems and resolving current ones. This might entail upgrading ethical standards, modifying development procedures, and giving the development and testing teams more training.

7.4 ADDRESSING BIAS AND DISCRIMINATION IN ALGORITHM DESIGN

In order to guarantee the moral and equitable use of artificial intelligence (AI) systems, prejudice and discrimination in algorithm design must be addressed. Particularly against historically oppressed groups based on gender, socioeconomic class, sexual orientation, or ethnicity, AI bias can have discriminatory effects. Prejudiced presumptions made during the model-development process, non-representative or erroneous training data, and algorithmic decision-making reproducing discriminatory social tendencies are some drivers of this bias [11–13]. In order to address these problems, machine-centric AI design must be used, and frameworks for evaluating bias impact must be created to increase public awareness of bias and its possible effects. This entails combining academic methods from computer science and philosophy with technological and managerial strategies, including building impartial databases, improving algorithm transparency, and setting up internal ethics governance [11, 13].

Finding and eliminating biases in AI systems is one of the main issues in combating algorithmic bias. Because algorithms are taught on previous data, they can unintentionally reinforce preexisting patterns of prejudice and bias. Even in the absence of overt human intent, this can result in the reproduction of social prejudices and the production of discriminatory consequences. In order to address this, it is critical to use best practices that direct the design, implementation, and monitoring stages of

AI systems, such as diversity-in-design, frequent bias audits, and the creation of bias impact statements.

Moreover, algorithmic bias may appear in a variety of forms. For example, algorithms have been discovered to display gender prejudice in online recruiting platforms. As an illustration of the practical effects of algorithmic prejudice on employment processes, Amazon stopped using a recruiting algorithm after identifying gender bias. At every level of the construction of an AI system, it is imperative to recognize these empirical truths and try to mitigate prejudices.

7.4.1 MITIGATING BIAS IN ALGORITHMIC DECISION-MAKING

Algorithms may accidentally harbor biases due to past prejudices and inequality reflected in the data used for training. Data pretreatment that is thorough and meticulous is one way to tackle this problem. In order to provide a representative and varied sample that considers a range of demographics and viewpoints, developers must evaluate and detect any potential biases in their training datasets. Furthermore, by balancing the dataset, methods like re-sampling, re-weighting, and augmentation can lessen the chance of enhancing or maintaining preexisting biases.

Algorithmic transparency, in addition to preprocessing, is essential for reducing bias. Developers must prioritize openness in their decision-making procedures, ensuring the algorithm's underlying mechanisms are comprehensible and accessible. Because of this transparency, external audits and inspections are made easier, enabling stakeholders to spot and correct biased practices. Additionally, it promotes user accountability for the algorithm's ethical ramifications. It helps to establish confidence among users by offering comprehensive documentation on the decision criteria and the effects of various features.

In development teams, diversity and cooperation are also essential for reducing prejudice. Teams that are made up of people with different experiences and backgrounds can work together to recognize and address any biases in algorithmic decision-making. Fostering an open atmosphere where team members are empowered to express concerns and suggest alternate viewpoints contributes to a more comprehensive analysis of potential biases during development.

7.4.2 REGULAR AUDITS AND REVIEWS FOR BIAS DETECTION

Regular audits are essential to ensure that ethical standards are being followed. They offer a methodology for assessing how well algorithms function in practical situations, which aids in identifying any underlying prejudices that could have accumulated over time. Development teams may proactively address developing concerns and preserve the integrity of their algorithms by regularly conducting audits. The stages, roles, resources, and results of the Bias Detection Audits process are outlined in Table 7.2 and are intended to help uncover and rectify biases in software algorithms.

As soon as prejudices are identified, remedial action must be taken quickly. Lessening the detected biases might entail improving the algorithm, changing the training set, or adding more security measures. Information regarding the audit results and next steps must be transparent to keep users and stakeholders trusting.

Frequent audits are a continuous commitment to developing ethical algorithms rather than a one-time occurrence. Development teams should continuously improve their auditing procedures, adjusting to changing user expectations, ethical standards, and application environment.

TABLE 7.2
Process of bias detection audits

Process Step	Responsibility	Tools/Methods	Output
Define Evaluation Metrics	Data Scientists, Ethicists	Define measurable criteria and benchmarks	Criteria for assessing bias in algorithms
Data Sampling	Data Scientists, Statisticians	Random sampling, stratified sampling	Diverse demographic data samples
Evaluate Output Disparities	Data Scientists, Analysts	Statistical analysis, visualization tools	Disparities and trends in algorithm output
User Feedback Integration	UX Researchers, Product Managers	Feedback collection mechanisms, surveys	User feedback on perceived biases
Comparative Analysis	Data Scientists, Developers	A/B testing, differential testing	Identification of changes and potential bias
Address Bias Discovered	Developers, Data Scientists	Algorithm adjustments, data retraining	Improved algorithm with reduced biases
Continuous Improvement	Ethicists, Project Managers	Periodic reviews, stakeholder consultations	Updated audit processes and guidelines

7.5 ETHICAL USE OF AI AND MACHINE LEARNING IN SOFTWARE DEVELOPMENT

The software development business has revolutionized by incorporating AI and machine learning (ML) techniques, which provide unparalleled prospects for innovation and efficiency enhancements. Nonetheless, when putting such cutting-edge technology into practice, ethical considerations must be taken into account.

Research shows that by automating monotonous operations like bug fixing and documentation, AI systems may significantly increase engineers' creativity and enhance the development process overall [14, 15]. Developers can find new patterns and information clusters by conducting structured assessments of large volumes of data using ML algorithms [14].

Although AI and ML have much potential, they also present special difficulties. For instance, ML models are intrinsically reductive rather than creative, as they frequently rely on structures that humans established. Furthermore, creating ML systems necessitates specific knowledge, dividing people into those who comprehend the underlying theory and those who use it in real-world applications [15].

By offering no-code AI platforms, recent advancements hope to democratize AI by enabling anyone with no programming skills to use AI and ML techniques [16]. More accessibility is made possible by this trend, but it also raises questions about the morality and caliber of AI products made by less skilled developers.

7.5.1   FAIRNESS AND ACCOUNTABILITY IN AI SYSTEMS

While accountability focuses on holding developers and AI systems accountable for their activities, achieving fairness entails minimizing biases and discriminatory results in AI decision-making. AI fairness aims to prevent unfair or biased treatment of people or groups on the basis of attributes like gender, race, or socioeconomic position. Training data, algorithms, and model outputs must be carefully examined to find and reduce possible biases.

Biased training data is one of the main causes of bias in AI systems. To guarantee a fair model, developers must carefully examine and preprocess training datasets, eliminating biases and aiming for diversity and representativeness. Fairness in algorithmic decision-making depends on transparency. Developers need to create algorithms that make sense and can be explained so that stakeholders and consumers can understand the process involved in making judgments. This openness also makes it easier to spot and address any inadvertent biases.

Creating accountability entails making developers responsible for ethical issues, creating accountability for the behavior of AI systems, and offering procedures for resolving unfavorable outcomes. This encourages confidence and trust in AI systems. Throughout the AI development process, developers should keep thorough documentation that covers the model architecture, data sources, and decision-making logic. This traceability makes responsibility easier and helps developers identify and address any problems.

Even if AI systems can automate many jobs, human monitoring should always be a part of the process. Developers must incorporate human assistance into AI systems when needed, particularly during crucial decision-making procedures, to maintain ethical standards and responsibility.

7.5.2   ETHICAL IMPLICATIONS OF DATA USAGE IN AI

While accountability focuses on holding developers and AI systems accountable for their activities, achieving fairness entails minimizing biases and discriminatory results in AI decision-making. We examine many important facets of accountability and justice in AI systems in this subsection: AI fairness aims to prevent unfair or biased treatment of people or groups on the basis of attributes like gender, race, or socioeconomic position. Training data, algorithms, and model outputs must be carefully examined to find and reduce possible biases.

Biased training data is one of the main causes of bias in AI systems. To guarantee a fair model, developers must carefully examine and preprocess training datasets, eliminating biases and aiming for diversity and representativeness.

Fairness in algorithmic decision-making depends on transparency. Developers need to create algorithms that make sense and can be explained so that stakeholders and consumers can understand the process involved in making judgments. This openness also makes it easier to spot and address any inadvertent biases.

Creating accountability entails making developers responsible for ethical issues, creating accountability for the behavior of AI systems, and offering procedures for resolving unfavorable outcomes. This encourages confidence and trust in AI systems.

Throughout the AI development process, developers should keep thorough documentation that covers the model architecture, data sources, and decision-making logic. This traceability makes responsibility easier and helps developers identify and address any problems.

Even if AI systems can automate many jobs, human monitoring should always be a part of the process. Developers must incorporate human assistance into AI systems when needed, particularly during crucial decision-making procedures, to maintain ethical standards and responsibility.

7.6 ENSURING ETHICAL STANDARDS IN SOFTWARE MAINTENANCE AND UPDATES

The continuous support and improvement of software to accommodate evolving needs and resolve problems that crop up after distribution is referred to as software maintenance. Updating software might include changing preexisting functionality, addressing security holes, or adding new features, making ethical issues even more crucial.

In order to ensure software maintenance adheres to ethical norms, businesses need to set up explicit policies and procedures for handling moral dilemmas that may come up during updates. In order to comprehend possible effects on user privacy, data security, and overall system integrity, this involves doing comprehensive impact assessments. It is equally crucial to openly and honestly communicate these ethical norms to users to build confidence and make sure that people are aware of how their data will be handled and how modifications to the system may affect how they interact with the program.

Another crucial component of maintaining ethical standards during software maintenance and upgrades is balancing security requirements and ethical concerns. While fixing security flaws is essential to protecting user data, it is also critical to carefully apply these security measures to prevent unwanted repercussions and privacy violations. Organizations must prioritize the ethical ramifications of security improvements, as they must carefully balance safeguarding user data with maintaining user liberty.

REFERENCES

1. Yip, C., N.-L.R. Han, and B.L. Sng, Legal and ethical issues in research. *Indian Journal of Anaesthesia*, 2016. 60(9): p. 684.
2. Biable, S. E., Garcia, N. M., Midekso, D., & Pombo, N, Ethical issues in software requirements engineering. *Software*, 2022. 1(1): pp. 31–52.
3. Varkey, B., Principles of clinical ethics and their application to practice. *Medical Principles and Practice*, 2021. 30(1): pp. 17–28.

4. Smith, C.S., Applying a systems oriented ethical decision making framework to mitigating social and structural determinants of health. *Frontiers in Oral Health*, 2023. 4: pp. 1–10.
5. Suarez, V. D., Marya, V., Weiss, M. J., & Cox, D, Examination of ethical decision-making models across disciplines: Common elements and application to the field of behavior analysis. *Behavior Analysis in Practice*, 2023. 16(3): pp. 657–671.
6. DeSimone, B.B., Curriculum design to promote the ethical decision-making competence of accelerated bachelor's degree nursing students. *Sage Open*, 2016. 6(1): p. 2158244016632285.
7. McDevitt, R., C. Giapponi, and C. Tromley, A model of ethical decision making: The integration of process and content. *Journal of Business Ethics*, 2007. 73: pp. 219–229.
8. Dacin, M. T., Harrison, J. S., Hess, D., Killian, S., & Roloff, J., Business versus ethics? Thoughts on the future of business ethics. *Journal of Business Ethics*, 2022. 180(3): pp. 863–877.
9. Martínez, C., A.G. Skeet, and P.M. Sasia, Managing organizational ethics: How ethics becomes pervasive within organizations. *Business Horizons*, 2021. 64(1): pp. 83–92.
10. Daradkeh, M., Navigating the complexity of entrepreneurial ethics: A systematic review and future research agenda. *Sustainability*, 2023. 15(14): p. 11099.
11. Belenguer, L., AI bias: Exploring discriminatory algorithmic decision-making models and the application of possible machine-centric solutions adapted from the pharmaceutical industry. *AI and Ethics*, 2022. 2(4): pp. 771–787.
12. Köchling, A. and M.C. Wehner, Discriminated by an algorithm: A systematic review of discrimination and fairness by algorithmic decision-making in the context of HR recruitment and HR development. *Business Research*, 2020. 13(3): pp. 795–848.
13. Chen, Z., Ethics and discrimination in artificial intelligence-enabled recruitment practices. *Humanities and Social Sciences Communications*, 2023. 10(1): pp. 1–12.
14. Alshammari, F.H., Trends in intelligent and AI-based software engineering processes: A deep learning-based software process model recommendation method. *Computational Intelligence and Neuroscience*, 2022. 2022: p. 1960684
15. Barenkamp, M., J. Rebstadt, and O. Thomas, Applications of AI in classical software engineering. AI *Perspectives*, 2020. 2(1): p. 1.
16. Sundberg, L. and J. Holmström, Democratizing artificial intelligence: How no-code AI can leverage machine learning operations. *Business Horizons*, 2023. 66(6): pp. 777–788

8 The Impact of Information Technology on Society and Well-Being

8.1 ETHICAL IMPLICATIONS OF TECHNOLOGICAL ADVANCEMENTS ON SOCIETY

The development of technology has led to a profusion of ethical questions that have a big impact on society. The exponential rise of digital technologies is a major worry when it comes to the application of technology in business because it has increased the potential for ethical dilemmas [1, 2]. Amidst the rapid evolution of technology, CEOs and corporations involved in its development and dissemination raise questions about their societal responsibility. Technology is involved in a wide range of other ethical issues beyond the imposition of political or ideological ideas, privacy intrusions, financial and bodily harm to individuals, and challenges to individual autonomy [2].

Emerging technologies bring privacy issues and moral conundrums to society that demand careful consideration. Among the most important concerns in big data and machine learning are privacy protection, personal information security, dangers of artificial intelligence (AI), transparency, and adherence to human values, fairness, safety, and accountability [3]. To optimize advantages and minimize drawbacks related to technological advancement, it is imperative to set ethical guidelines for creating and disseminating novel technologies. Because of these ethical aspects of technology, preventative procedures like ethical technology assessment (eTA) are necessary to address any harmful effects of technology early on [4].

The growing use of AI in work settings raises important ethical questions about meaningful human employment. Even while AI technologies can potentially increase productivity, any changes that include assigning less important work to individuals need to be carefully considered. This is because meaningful labor has an ethical value. After all, it contributes to the prosperity and flourishing of humanity. Even though the ethical ramifications of implementing AI are widely acknowledged, further academic research is necessary to examine the precise ways AI affects meaningful labor. This knowledge gap emphasizes the need for additional study to fully understand and address the evolving ethical landscape that technology innovations are creating [5].

DOI: 10.1201/9781003584452-8

8.1.1 Privacy Concerns in the Digital Age

Technology has brought about previously unheard-of levels of efficacy and convenience, but it has also greatly increased privacy concerns, which are pervasive in modern life. People have to deal with the problem of protecting their privacy in a world where everything is connected. This problem arises in an era where people frequently view personal information as a commodity. Several digital businesses, social media networks, and online platforms regularly gather a lot of personally identifiable information—often without the user's knowledge or consent. Given this fact, questions about ethics and the proper handling, storage, and use of sensitive data have been raised.

One of the main concerns when protecting personal data is getting informed permission. Users may unintentionally allow illegal access to confidential information through seemingly casual online conversations. In order for users to make educated decisions about what information they disclose and how it is used, privacy policies need to be easily understood and available before ethical considerations are taken into account.

The gathering and examining of individually identifiable data for various objectives, such as targeted advertising or algorithmic decision-making, could result in unanticipated consequences. Several topics need to be thoroughly examined, including the possibility of discrimination, the moral implications of data profiling, and the creation of digital profiles that might not be true to life.

Laws and regulations—such as the General Data Protection Regulation (GDPR)—have been implemented to allay certain privacy worries by giving people more control over their data. Nevertheless, because of the global reach of the digital environment, continuous efforts are required to strengthen and harmonize privacy protections across legal frameworks. Table 8.1 compares and contrasts the implementation of key features of privacy policies.

TABLE 8.1
Privacy policy features comparison

Features	Accessible via Mobile Apps	Clear Language	Detailed Data Categories	Clear Consent Options	Email Updates	GDPR Compliance
Accessibility	✓					
Comprehensibility		✓				
Transparency			✓			
Consent Mechanisms				✓		
Updates and Notifications					✓	
Compliance with Regulations						✓

8.1.2   Surveillance and Ethical Boundaries

In terms of surveillance, a balanced approach requires preserving individual privacy given the need to guarantee public safety, regardless of the means of operation (official government agency, private firms, or developing technology). In this field, ethical discourse is concerned with defining the boundaries of surveillance activities to guarantee that they adhere to democratic values and protect individual rights.

Surveillance technologies present several ethical questions regarding the transparency, extent, and purpose of monitoring. Concerns have been raised about how much privacy, civil rights, and individual liberty are violated by surveillance. One of the biggest challenges is finding a balance that allows for the protection of public safety while simultaneously acknowledging the moral bounds of individual liberty.

To preserve the integrity of moral surveillance practices, monitoring operation supervisors need to be open and accountable in performing their tasks. Commercial and public entities must communicate the goals, workings, and constraints of surveillance technologies honestly and openly. Systems that hold surveillance operators accountable for infractions must be implemented according to ethical principles. This is carried out to ensure that the people use their authority with diligence.

Surveillance poses ethical questions about civil liberties and human rights that need to be addressed. Due to the possibility of bias, profiling, and unauthorized access, it is crucial to have ethical safeguards in place. For ethical grounds, legislative frameworks that protect people's freedoms of assembly, speech, and movement without unjustified intervention are essential. These guidelines are anticipated to protect people from unapproved surveillance.

Respecting ethical issues means holding open discussions, giving communities a say in using and regulating surveillance technology, and getting informed consent from the general public. By using this participative method, it is possible to get around adopting surveillance techniques that could be viewed as too intrusive.

Insofar as possible, ethical monitoring practices aim to reduce bias and discrimination. This entails eliminating disproportionate penalties for marginalized groups, treating diverse communities fairly, and addressing algorithmic biases in monitoring systems. In order to reduce the likelihood of unexpected outcomes, ethical frameworks should fervently support ongoing assessment and improvement.

8.1.3   Data Security and Cybersecurity Ethics

Data security concerns are not limited by technology; they are entwined with moral precepts that govern how sensitive information should be handled. Data security ethics cover a wide range of topics, one of which is the protection of sensitive company information as well as personally identifiable information. It is the ethical duty of organizations to take precautions against cyberattacks, illegal access, and data invasions by putting strong security measures in place. The basis of this requirement is the recognition of the possible harm that could arise from the breach of sensitive data within an organization.

People need access to complete and accurate information about how their data is being used, and they must be able to give their informed consent. Increasing consumer trust in companies can be accomplished by implementing open and honest privacy policies, easily navigable terms of service, and unambiguous communication regarding data-handling protocols. Respecting people's right to privacy and autonomy is essential to upholding ethical standards in data security.

The concept of cybersecurity ethics includes proactive risk-mitigation tactics in addition to reactive ones. It is the responsibility of organizations to not only address threats that have been detected but also to foresee and prepare for new threats. This ethical duty comprises several components, including investing in cutting-edge security technologies, proactively patching vulnerabilities, and being conscious of how cyber dangers are always changing. The prevention of harm is emphasized by ethical cybersecurity practices, which also foster a culture of continuous improvement and adaptation to emerging threats.

It is essential to guarantee prompt and open communication with the affected individuals. Furthermore, committing to enhancing security standards and putting corrective actions in place is critical. When reacting to data intrusions, ethical methods put accountability first, own up to mistakes, and work together to prevent similar incidents in the future. This dedication to correction and transparency accomplishes that goal by indicating the organization's moral position on data protection and regaining the trust of stakeholders.

As technology progresses, ethical data security becomes increasingly important to balance the delicate line between encouraging innovation and honoring moral commitments. Encouraging the development of cutting-edge technologies must go hand in hand with an ethical approach that prioritizes protecting user data and maintaining the overall integrity of digital infrastructure. By using ethical cybersecurity methods, a safe and sustainable digital ecosystem can be established. The welfare of businesses and individuals, as well as the advancement of technology, are all benefited by this environment.

8.1.4 ACCOUNTABILITY IN AI AND MACHINE LEARNING SYSTEMS

The functioning of AI and machine learning systems is supported by sophisticated algorithms that can learn and adjust. Moral quandaries arise when forecasts or conclusions are derived from analyzing enormous datasets in which the decision-making processes are not transparent. The complex functionality of these systems can pose difficulties in attributing responsibility in the event of errors or biased behavior.

It is the responsibility of businesses and developers to disclose the algorithms and decision-making processes utilized in the generation of visible content. This will enhance stakeholders' clarity and comprehension of the decision-making process. Facilitating external analysis and fostering accountability for the outcomes of AI- and ML-related processes, transparency contributes to the development of confidence.

Biased AI systems, whether deliberately designed or pervasive, give rise to ethical concerns. To ensure accountability, a concerted effort must be undertaken to detect and rectify any biases in the training data and algorithms. Ethical standards should promote the ongoing development, auditing, and monitoring of AI models to mitigate the potential for biased outcomes. Responsible development strategies aim to balance the imperative to ensure the ethical integrity of AI systems and the requirement for progress.

Ethical frameworks must be implemented to establish clear obligations of organizations, regulatory bodies, and developers concerning the mitigation of harm, resolution of concerns, and accountability for the consequences of choices generated by AI, ML, and associated technologies.

Accountability is a fluid concept within the domain of AI that necessitates continuous evaluation and improvement. Compliance with ethical concerns necessitates acquiring knowledge from errors, refining conceptual frameworks, and integrating input from alternative perspectives. The perpetual progression of responsible AI development is predicated on resolving emergent ethical concerns and adapting frameworks to align with evolving societal standards.

8.2 DIGITAL DIVIDE AND ETHICAL CONSIDERATIONS OF ACCESSIBILITY

Various factors, including digital literacy, economic opportunity, political participation, and technological access, can dissect the divide. It may be more challenging for individuals lacking access to high-speed internet to participate in political and civic engagements; this serves as a singular example of the pervasive consequences that can result from the digital divide [6].

Concerns abound regarding the "digital divide" in eHealth about accessibility for individuals with disabilities. There is a possibility that individuals with disabilities may face difficulties when endeavoring to utilize eHealth services as a result of design components that are insufficiently customized to accommodate their specific requirements [7].

Pre-existing disparities in educational opportunities have been brought to light by the transition to online learning, which has also brought to light inequities in the availability of digital infrastructure. To ensure equitable access to study materials and educational resources for all pupils, it is critical to develop inclusive methodologies that consider the contrasting conditions of urban and rural environments, as exemplified by the extensive utilization of mobile internet data for online learning by students. Additional guidance from instructors after online courses conclude may improve students' academic performance in navigating the challenges the digital age introduces to education [8].

8.2.1 Socioeconomic Factors and Access Disparities

The "digital divide" concept denotes the disparity in the availability and utilization of information and communication technologies. Socioeconomic disparities are intricately connected to this incongruity and contribute to worldwide disruptions that affect the lives of individuals and communities. Notably, individuals' social circumstances substantially influence the extent to which they can exploit technological advancements' opportunities. This phenomenon brings attention to the ethical concerns that pertain to digital accessibility. The growing disparity in global knowledge is illustrated in Figure 8.1, which highlights the effects on people, communities, and nations.

Numerous factors, including an individual's educational attainment, earning potential, and career options, substantially impact their ability to acquire and employ digital technologies. Given the growing significance of technology in civic participation, employment, and education, it is critical to address these disparities to prevent the consolidation of inequality.

Accessibility to technology is contingent not only on the presence of physical infrastructure but also on an individual''s proficiency in its operation. An element contributing to digital literacy disparities is the widespread occurrence of educational inequality. Individuals from lesser socioeconomic strata may need more competencies to navigate the digital realm effectively under these circumstances. In order to bridge this divide, it will be necessary to incorporate ethical considerations into the design of inclusive educational initiatives and programs that seek to equip all individuals with the skills and knowledge necessary to participate completely in the digital age.

Financial constraints may pose a substantial obstacle for individuals due to the substantial costs associated with technology, including equipment acquisition and internet connectivity. Ethical considerations ought to be incorporated into efforts to rectify access disparities, including exploring strategies to reduce technology costs, endorsing legislation that subsidizes critical digital services, and promoting public-private partnerships to ensure universal economic accessibility.

Recognizing the necessity of digital inclusion for social justice underscores the ethical dimension associated with eliminating obstacles to access. Consideration must be given to the broader ramifications of social justice when undertaking initiatives to bridge the digital divide. Beyond facilitating entry, these initiatives should prioritize empowering underrepresented individuals and groups. A method for assuming an ethical posture is to comprehend the intersectionality of variables such as race, gender, and geography. This ensures that solutions are all-encompassing and adaptable to meet the diverse requirements of users.

In the legislative process, the active participation of governments, business stakeholders, and advocacy organizations is vital for developing legislation that effectively tackles access gaps. Developing digital literacy programs, promoting affordable digital services, and establishing an inclusive digital infrastructure ought to be the primary aims of policy initiatives. Consistently emphasizing the moral obligation to bridge the digital divide is crucial for establishing a society wherein all individuals can participate, offer input, and benefit from the possibilities presented by information technology.

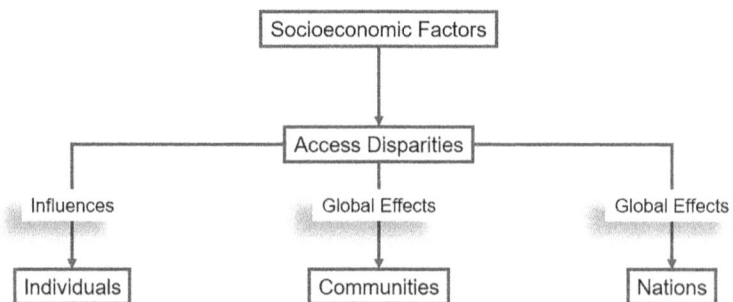

FIGURE 8.1   Socioeconomic disparities and global impact

8.2.2 INCLUSIVE DESIGN AND ETHICAL TECHNOLOGY ACCESSIBILITY

Developing solutions that consider geographic variances, cultural diversity, and economic limits is vital to guarantee the accessibility of technology that adheres to ethical standards. Inclusive design encourages the moral use of technology by recognizing and removing obstacles to access, benefiting a wider range of people.

The accessibility of ethical technology incorporates the ideals of inclusive design. The concepts listed above prioritize the development of products and services that are useful for people with varying skill levels and characteristics. This part covers the creation of user-friendly user interfaces, the availability of various information access points, and the guarantee of assistive technology compatibility. In order to achieve the goal of providing technology access to everyone, it is necessary to create an atmosphere that permits everyone to engage in the digital era fully.

A key element of ethical technology accessibility is digital inclusion, which ensures that everyone has the necessary resources, knowledge, and abilities to access and use information and communication technologies. Ethical approaches include legislation and programs that support digital literacy, offering training, and raising public understanding of technology's benefits. It is essential to build a relationship through assistance and education in order to promote a technology environment that is both more morally responsible and inclusive.

8.3 ETHICAL CHALLENGES IN TECHNOLOGICAL INNOVATIONS AND DISRUPTIONS

The speed at which technology develops has led to a conundrum whereby advances in AI could spur innovation. However, not accompanied by appropriate governance, they could exacerbate power imbalances. To ensure the appropriate development and application of technology, a system that combines government control and self-regulation is necessary. This is necessary to strike a healthy balance between the need for innovation and moral obligations. To reduce the ethical concerns connected with their breakthroughs, companies like Facebook, Google, and Microsoft are under increasing pressure to make their data management rules more publicly overseen [9].

Ethical technology assessment (eTA) has been proposed as a conceptual framework to address developing technologies' ethical, legal, and societal implications. These frameworks are used throughout development as early warning systems for possible unethical consequences. These frameworks aim to guarantee that technological advancements align with moral principles. The writers emphasize how crucial it is to take gender diversity and privacy into account from the outset when creating new information technologies. It is imperative to incorporate ethical considerations into the technological development process to proactively identify and address any ethical issues that may arise with the broad adoption of novel technologies [4].

8.3.1 DISRUPTION AND JOB DISPLACEMENT: ETHICAL EMPLOYMENT CONCERNS

Job displacement can worsen already-existing socioeconomic inequities, especially when certain firms or demographic groups are disproportionately affected. Moral frameworks must consider the possibility of higher unemployment rates among particular groups and ensure that the advantages of technological advancement do not go to a small minority. Planning and carrying out programs that support diversity and inclusion and provide targeted aid to the impacted groups to allay these ethical worries is essential.

When it comes to solving the issue of job displacement, an ethical framework demands that technology innovation that is focused on people be given priority. Throughout the development process, it is crucial to consider how technology improvements may affect employment to accomplish this goal. Above all, moral technology should seek to enhance rather than completely replace human talents. This will encourage human-robotic collaboration and result in the creation of more inclusive and sustainable work environments.

Investing in comprehensive reskilling and retraining initiatives is crucial for governments and businesses to successfully address the ethical challenges related to job displacement. Offering instructional programs that are not only easily accessible but also economical and productive can help staff members adapt to rapidly changing technology settings. Creating environments that support lifelong learning and guarantee that no one is left behind in the face of technological advancements falls under the category of ethical considerations.

Businesses that are at the forefront of technology innovation have moral obligations when it comes to handling job displacement. Corporate social responsibility programs should prioritize financial gains and the welfare of employees and the community. This includes providing reasonable and equitable severance benefits, being open and honest about any planned changes to the workforce, and taking the initiative to participate in community development projects.

8.3.2 Cultural and Social Impact of Technological Innovations

Technological advances constitute its foundation and profoundly impact various domains, including economic systems, community structures, human relationships, and language (Figure 8.2).

Social networking platforms have become pivotal in facilitating interpersonal connections by transcending geographical barriers. Notwithstanding this, the effects on interpersonal communication, compassion, and the quality of connections remain a topic of continuous discourse. Ethical implications emerge concerning the capacity of technology to augment or diminish the caliber of social connections.

The proliferation of automation engenders ethical concerns regarding job displacement, retraining, and the socioeconomic ramifications of technological unemployment. A crucial societal challenge is striking a balance between the efficiency gains of technology and ethical considerations about equitable access to employment opportunities. Ethical frameworks must incorporate provisions that guard against the potential for technology to worsen social inequalities and establish principles that govern responsible business conduct.

The disruption or extinction of conventional modes of knowledge, languages, and cultural customs may result from accelerating technological progress. Technological innovations give rise to ethical considerations that necessitate the delicate equilibrium between embracing advancements and safeguarding cultural diversity. Efforts to archive, document, and safeguard cultural heritage in a digital format are crucial for ensuring the ethical preservation of cultural wealth.

Pre-existing disparities can be exacerbated by an uneven allocation of technological resources, giving rise to "digital divides" that marginalize specific demographic groups. A dedication to confronting these disparities through the promotion of inclusive technological development, the guarantee of accessibility for all, and the active pursuit of bridging socioeconomic gaps is mandated by ethical considerations.

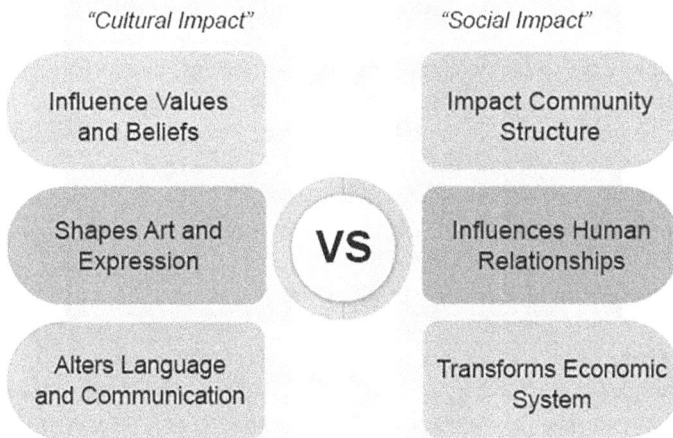

"Cultural Impact"   "Social Impact"

Influence Values and Beliefs

Impact Community Structure

Shapes Art and Expression   **VS**   Influences Human Relationships

Alters Language and Communication

Transforms Economic System

FIGURE 8.2   Dynamics of technological innovations

8.4   ETHICAL USE OF TECHNOLOGY FOR SOCIAL GOOD AND SUSTAINABLE DEVELOPMENT

Ethical considerations arise when intelligent information technologies are employed to help achieve the United Nations' Sustainable Development Goals (SDGs) [10]. The previously outlined challenges are connected to the dependability of the data, the availability of technological resources, and the cost implications of deploying these systems. Organizations that actively develop technology must navigate specific ethical quandaries to ensure that their breakthroughs uphold moral ideals and benefit society.

Continuous inquiries are into the relationship between large technology corporations and society's long-term viability. This stresses the importance of successfully developing an ethical framework to guide scientific research toward long-term outcomes [11]. This framework seeks to create a harmonic balance between ecological, social, and economic objectives while safeguarding individual liberties, rights, and democratic values, given the potential impact of technological innovation on labor markets, privacy, and social cohesion.

Engineers must use a holistic paradigm that considers morality when designing and managing complex socio-technical systems. This holistic engineering paradigm prioritizes social well-being, ethical coherence, and field harmonization. This is done to effectively address the complex issues outlined in the SDGs.

Addressing ethical flaws in technology development for the benefit of society is critical to promoting sustainable development and ensuring that technological discoveries adhere to ethical values [12]. This will make it easier to reconcile ethical standards with technological advances. The evident contribution of AI to achieving SDGs emphasizes the ethical challenges surrounding AI development and implementation [13].

8.4.1 LEVERAGING TECHNOLOGY FOR SOCIAL IMPACT

There are new ways to tackle complex societal problems, thanks to digital innovation. One example is using big data analytics to help make informed decisions, and another is the introduction of mobile apps to encourage community involvement. Organizations and people are investigating fresh ways to apply tech-driven solutions to make social efforts and projects more effective.

People on the margins now have access to resources, knowledge, and opportunities previously unavailable because of digital platforms and connections. Because of this, these people can take advantage of advantageous situations. The societal gaps can be filled, and the benefits of technological advancement can be shared fairly among everyone if this inclusive approach is implemented.

Two examples of how technical progress greatly aids environmental sustainability are using renewable energy sources and introducing smart farming techniques. Verifying that technological interventions align with SDGs and contribute positively to environmental conservation is an ethical factor relevant to this discourse.

Corporate social responsibility (CSR) programs are becoming more common among tech corporations. These corporations have realized how much they can shape the future, and this is the result. Beyond making money, these companies are committed to doing good in the world through charitable work, community development, and ethical business practices. An essential part of the company's CSR strategy is promoting technology's positive social impact by highlighting the positive contributions it can bring to society.

By implementing community-driven initiatives, digital literacy programs, and technology-accessible infrastructure, individuals can actively contribute to the betterment of their communities and their personal satisfaction. Encouraging the active involvement of local communities in technology development is often an essential requirement for the successful application of technology for societal benefit. By embracing this grassroots approach, technology can be turned from a tool of dependence to a tool of empowerment.

8.4.2 SUSTAINABLE TECHNOLOGY: ENVIRONMENTAL AND ECOLOGICAL ETHICS

Utilizing technology to its full capacity to benefit the environment, reduce negative impacts, and advance sustainability is ethically imperative. A comprehensive evaluation of the environmental effects of technology developments is necessary for ethical reasons as they alter many businesses. Product life cycle assessments consider resource extraction, manufacturing procedures, and disposal strategies. In order to promote responsible innovation that reduces damage to ecosystems and biodiversity, ethical and ecological impact evaluations seek to identify and mitigate negative outcomes.

The goal of sustainable technology is to promote social and economic development while simultaneously reducing the negative impact on the environment. Responsible technology development in renewable energy, circular economies, and resource efficiency is guided by ethical principles to ensure that environmental integrity is not compromised.

People are increasingly expecting IT companies to do the right thing for the environment, and that responsibility goes beyond just making a profit. Transparency, accountability, and reducing carbon footprints are ethical considerations in sustainable technology. A commitment to ecological ethics should guide CSR efforts in the technology sector, showing that the sector values sustainability and cares about the environment.

8.4.3   CSR IN THE TECH INDUSTRY

A common focus of CSR programs at technology businesses is creating and implementing technological solutions to pressing environmental and social problems. The ability of technology to improve society by creating useful applications in areas like healthcare, education, and environmental protection is demonstrated by such attempts.

CSR in the IT industry has evolved, with a focus on sustainable practices and the creation and use of eco-friendly technologies being at the forefront. By reducing technological waste, investing in renewable energy sources to power operations, and implementing energy-efficient infrastructure, a corporation can demonstrate its commitment to ecological ethics.

Technology businesses are starting to acknowledge the importance of ethical considerations during production and sourcing due to the growing complexity of global supply networks. Components of CSR programs include reducing environmental effects across the supply chain, ensuring ethical labor practices, and consistently exploiting resources. We can encourage ethical behavior and hold IT corporations and their suppliers to account if we are open and honest about these practices.

8.5   IMPACT OF IT ON MENTAL HEALTH AND WELL-BEING

The effects of social media on people's health might be mixed [14, 15]. On the one hand, social media can help people feel more connected, confident, and part of a community. They allow people to connect, talk about health issues, and offer emotional support, which is especially important at difficult times like the COVID-19 epidemic [15]. A decline in psychological health, increased loneliness, FOMO, and dissatisfaction with life have all been linked to heavy social media use [15, 16].

There are benefits to social media, like connecting with friends and making new acquaintances. However, there are also risks, such as having an unhealthy obsession with the internet, being a victim of cyberbullying, and experiencing bad mood swings [15, 16]. The potential benefits and drawbacks of technology workplace improvements on mental health have recently been brought to light in research. These developments might affect individuals' stress levels, work-life balance, and general health, including more connections, remote work arrangements, and digital communication tools [17].

8.5.1   Screen Time and Digital Addiction

Due to the ubiquitous digitization of displays, modern folks spend significant time immersed in digital domains. The ubiquitous usage of displays has given rise to moral worries about the potential unintended effects of technology on human life.

Anxiety, despair, and trouble falling asleep are just a few of the negative outcomes linked to spending too much time in front of screens. Because of the potential negative impact on users' mental health, apps and platforms that promote heavy usage give rise to ethical problems. Ethical considerations of mental health must take precedence when seeking user participation in responsible technology creation.

Addiction to digital media can have devastating effects, and young people are especially vulnerable to these effects. Ethical questions arise when considering who protects vulnerable populations: parents or individuals creating innovative technologies. Promoting healthy digital habits and balancing instructional and recreational screen time are ethical responsibilities.

A "digital detox" is a perfect example of the moral conundrum people face when trying to balance their responsibilities with technology and their independence. People must have complete control over how much time they spend on digital devices and how they engage with these platforms. Not only is it important to design user interfaces and apps with people's well-being in mind, but it is also important to give people the tools to track and manage their digital usage.

Because of their pivotal position in creating digital experiences, information technology (IT) corporations have an ethical obligation to consider how their products will affect user behavior. Initiatives that promote digital well-being and encourage open discussion about the risks associated with screen addiction are becoming more and more justified from an ethical standpoint. IT businesses should be commended for their CSR actions that address digital addiction, which is a significant public health risk.

8.5.2   Mental Health Apps

There has been a dramatic increase in the availability of resources meant to promote psychological health thanks to the meteoric rise of mental health applications in the past few years. However, serious ethical questions regarding implementing such technologies must be thoroughly investigated. Many mental health apps secretly record users' innermost feelings, ideas, and actions. Ensuring the confidentiality and integrity of the obtained information is crucial in ethical app development. Powerful encryption mechanisms, explicit user permission processes, and rigorous adherence to data protection standards can achieve this.

In order to build trust with users and stop harmful or false content from getting out, the app needs to be transparent about its techniques, sources of information, and validation studies.

Developers should adequately explain the app's goal, features, and possible results so users can make educated decisions regarding their mental health care. We respect user autonomy by avoiding coercive aspects in the app UI, allowing for customization and termination options, and avoiding excessive influence.

Creators who prioritize ethics make sure that resources for mental health are presented in a way that is both accepting and respectful of different cultures. To further reduce the possibility of incorrect diagnoses or worsening of mental health disorders, app algorithms should undergo extensive testing. Developers are responsible for designing accessible interfaces that can work for people with different backgrounds, language skills, and physical or mental limitations. Adhering to universal design principles can amplify these technologies' ethical influence. To build and release mental health apps responsibly, it is important to consider the ethical factors listed in Table 8.2.

TABLE 8.2
Ethical issues in the development of mental health applications

Ethical Considerations	Techniques	Sources of Information	Validation Studies
Confidentiality and Integrity	Robust encryption data protection standards	Secure data storage, encryption mechanisms	Compliance with data protection laws
Transparency	Clear communication, disclosure of methods	Transparency reports, disclosure statements	Validation studies, peer-reviewed research
Informed Consent	Explicit permission processes	Consent forms, user agreements	User feedback informed consent studies
User Autonomy	Non-coercive UI design	Opt-in features, termination options	User satisfaction surveys
Cultural Sensitivity	Respectful presentation of resources	Cultural competency training, diverse content	Cultural validation studies
Accuracy and Safety	Extensive algorithm testing	Clinical trials, benchmarking against standards	Accuracy assessments, clinical validation
Accessibility	Universal design principles	Accessibility guidelines, user testing	Accessibility audits, usability studies

8.5.3 BALANCING CONNECTIVITY AND EMOTIONAL WELL-BEING

The advent of cell phones, social media platforms, and instant communication technologies has undeniably bolstered the velocity and scope of information dissemination. Nevertheless, the continuous state of being connected has given rise to ethical considerations about its potential ramifications on emotional welfare. Engaging in a continuous pursuit of updates, notifications, and digital engagements can potentially result in an overwhelming amount of information, which can lead to increased levels of tension, anxiety, and, in severe instances, digital addiction.

The ethical obligation resides in cultivating a harmonious equilibrium between the benefits of connectivity and safeguarding emotional welfare. This necessitates a comprehensive understanding of individual requirements, psychological limitations, and the acknowledgment of technology's function in enhancing rather than supplanting authentic human relationships. The design of digital interfaces and apps must consider ethical considerations, particularly in user experience, with a specific emphasis on mental health and emotional resilience.

8.6 ETHICAL RESPONSIBILITIES IN PROMOTING DIGITAL INCLUSION AND EQUALITY

It is the moral obligation to create policies that are accessible to all groups and meet their unique requirements, taking into account variables like income level, geography, and cultural norms. To cultivate an ethical ecosystem that promotes equal opportunity for all, regulatory frameworks should prohibit discriminatory acts and increase accessibility simultaneously.

Equal access to technology-related education is key to closing the digital gap. Formal education and lifelong learning and skill enhancement programs fall under this category. In order to ensure that everyone in society can benefit from digital breakthroughs, it is ethically vital that these programs consider different learning styles and accommodate people with different levels of technological ability. Figure 8.3 shows an ethical framework for digital inclusion and equality that addresses accessibility, representation, universal design, affordability, various perspectives, and prejudice reduction.

The failure to properly plan algorithms and technology has the potential to make existing inequities worse. In order to prevent algorithms from unintentionally reinforcing societal preconceptions, it is ethically necessary to identify and remove biases from AI systems. Developing and deploying technology should be guided by a commitment to diversity and fairness, with continuing evaluations to prevent unforeseen impacts.

A more equitable technology sector is the result of efforts to increase the representation of people from all demographics, including but not limited to gender, race, and socioeconomic position. Mentorship programs, inclusive hiring policies, and other efforts to foster a friendly and inclusive work environment are morally necessary. Tech companies can improve their problem-solving and creativity while meeting their ethical commitments by actively seeking out and hiring people from various backgrounds and experiences.

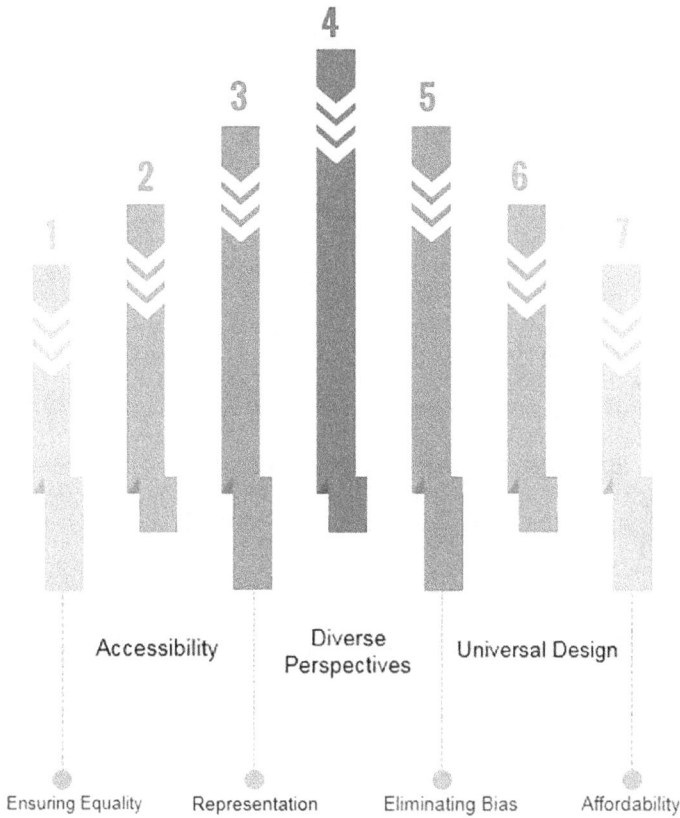

FIGURE 8.3 Ethical factors for digital inclusion

REFERENCES

1. Herschel, R.T. and P.H. Andrews, Ethical implications of technological advances on business communication. *The Journal of Business Communication (1973)*, 1997. 34(2): pp. 160–170.
2. Martin, K., K. Shilton, and J.E. Smith, *Business and the ethical implications of technology: Introduction to the symposium*. in Kirsten Martin, Katie Shilton, and Jeffery Smith (eds) *Business and the ethical implications of technology.*2022: Springer. pp. 1–11.
3. Dhirani, L. L., Mukhtiar, N., Chowdhry, B. S., & Newe, T, Ethical dilemmas and privacy issues in emerging technologies: A review. *Sensors*, 2023. 23(3): p. 1151.
4. Kendal, E., Ethical, legal and social implications of emerging technology (ELSIET) symposium. *Journal of Bioethical Inquiry*, 2022. 19(3): pp. 363–370.
5. Bankins, S. and P. Formosa, The ethical implications of artificial intelligence (AI) for meaningful work. *Journal of Business Ethics*, 2023. 185(4): pp. 725–740.
6. Sanders, C.K. and E. Scanlon, The digital divide is a human rights issue: Advancing social inclusion through social work advocacy. *Journal of Human Rights and Social Work*, 2021. 6(2): pp. 130–143.

7. Pettersson, L., Johansson, S., Demmelmaier, I., & Gustavsson, C., Disability digital divide: Survey of accessibility of eHealth services as perceived by people with and without impairment. *BMC Public Health*, 2023. 23(1): p. 181.

8. Jafar, K., K. Ananthpur, and L. Venkatachalam, Digital divide and access to online education: New evidence from Tamil Nadu, India. *Journal of Social and Economic Development*, 2023. 25(2): pp. 313–333.

9. De Cremer, D. and G. Kasparov, The ethics of technology innovation: A double-edged sword? *AI and Ethics*, 2022. 2(3): pp. 533–537.

10. Ryan, M., Antoniou, J., Brooks, L., Jiya, T., Macnish, K., & Stahl, B., The ethical balance of using smart information systems for promoting the United Nations' Sustainable Development Goals. *Sustainability*, 2020. 12(12): p. 4826.

11. Arogyaswamy, B., Big tech and societal sustainability: An ethical framework. *AI & Society*, 2020. 35(4): pp. 829–840.

12. Powell, A. B., Ustek-Spilda, F., Lehuedé, S., & Shklovski, I., Addressing ethical gaps in 'Technology for Good': Foregrounding care and capabilities. *Big Data & Society*, 2022. 9(2): p. 20539517221113774.

13. Vinuesa, R., Azizpour, H., Leite, I., Balaam, M., Dignum, V., Domisch, S., Felländer, A., Langhans, S.D., Tegmark, M. and Fuso Nerini, F., The role of artificial intelligence in achieving the Sustainable Development Goals. *Nature Communications*, 2020. 11(1): pp. 1–10.

14. Karim, F., Oyewande, A.A., Abdalla, L.F., Ehsanullah, R.C. and Khan, S., Social media use and its connection to mental health: A systematic review. *Cureus*, 2020. 12(6).

15. Zsila, Á. and M.E.S. Reyes, Pros & cons: Impacts of social media on mental health. *BMC Psychology*, 2023. 11(1): p. 201.

16. Naslund, J.A., Bondre, A., Torous, J. and Aschbrenner, K.A., Social media and mental health: Benefits, risks, and opportunities for research and practice. *Journal of Technology in Behavioral Science*, 2020. 5: pp. 245–257.

17. Johnson, A., Dey, S., Nguyen, H., Groth, M., Joyce, S., Tan, L., Glozier, N. and Harvey, S.B., A review and agenda for examining how technology-driven changes at work will impact workplace mental health and employee well-being. *Australian Journal of Management*, 2020. 45(3): pp. 402–424.

9 Social Networking and Online Communities
Ethical Engagement

9.1 ETHICAL CONSIDERATIONS IN BUILDING ONLINE COMMUNITIES

Privacy, informed consent, and online safety are just a few ethical factors to consider while constructing online communities. The necessity to take into account the specific circumstances of online research, the possibility of damage, and the porous borders between the public and private realms are among the difficulties that researchers have uncovered in this field [1, 2].

The assumption of privacy in digital environments is a critical one. Legally, things are more complicated, even though social media users may think their data is private. Since users have already agreed to the terms of service for social media platforms, courts in the United States have determined that users do not reasonably expect to protect the privacy of the information they share on these platforms [3]. On the other hand, there still needs to be a clear consensus on how much privacy users should anticipate or when posts become public knowledge. When constructing virtual communities, informed consent is an additional important ethical factor to think about. Researchers should think about ways to get users' informed consent if they require it, especially if the data will be made public. According to one study, incentives for involvement can be morally tricky since they can ruin people's anonymity and lead to low recruitment rates when done online [4].

Researchers have established several ethical frameworks and principles to deal with these issues. In response to the specific difficulties of conducting research online, several organizations have revised their standards, including the Association of Internet Researchers (AoIR), the British Psychological Society, and the British Society of Criminology [2]. Researchers have methodological and ethical issues when engaging people online, especially when tracing participants and engaging adolescents [4, 5].

9.1.1 FOSTERING INCLUSIVITY AND DIVERSITY

Members of diverse online communities were likelier to develop novel solutions to problems. However, getting there calls for focused energy and persistence. To be inclusive, one must do more than make room for other perspectives; one must genuinely welcome and appreciate people from all walks of life, irrespective of age, socioeconomic class, race, ethnicity, gender, or sexual orientation. To build an inclusive culture, people must try to comprehend and resolve the specific problems encountered by disadvantaged groups.

It is critical to have clear rules and regulations that encourage courteous and welcoming conduct to cultivate diversity and inclusion in online communities. Part of this is making sure that victims of harassment have access to resources and support, as well as enforcing stringent regulations against harassment. In order to ensure that excluded communities have equal opportunity to engage and contribute, it is crucial to seek out and amplify their views actively. If hidden biases exist in the community, it is critical to identify them and work to eliminate them. Unconscious prejudices can greatly influence marginalized communities' experiences in online environments. So, to make sure that everyone is cognizant of and trying to overcome these biases, educating and teaching community members and moderators about unconscious bias is crucial.

9.1.2 TRANSPARENCY IN COMMUNITY MANAGEMENT

Empirical evidence suggests that transparency is both a legal mandate and a useful instrument for public managers to engage with external stakeholders. Increased accountability, trust, and the prevention of corruption are all facilitated by transparency. Nonetheless, some contend that openness can exacerbate external intervention and erode confidence. Examining whether transparency is a prerequisite for developing public value and differentiating between the many forms of transparency organizations can seek are crucial steps toward overcoming this dichotomy [6].

Since it is a crucial accountability component, transparency is especially significant in non-profit organizations. Transparency is a critical component of good governance and is frequently justified by managerial accountability [7]. Communication, explicit policies, participation, accountability, decision-making, data transparency, and documentation are all components of transparent community management (Figure 9.1).

Transparency and accountability, however, rely on committee and community power structures and cultures, and there might not be official channels for people to file complaints or seek information from water committees. International non-governmental organizations may offer training or assist in facilitating channels for accountability and transparency [8].

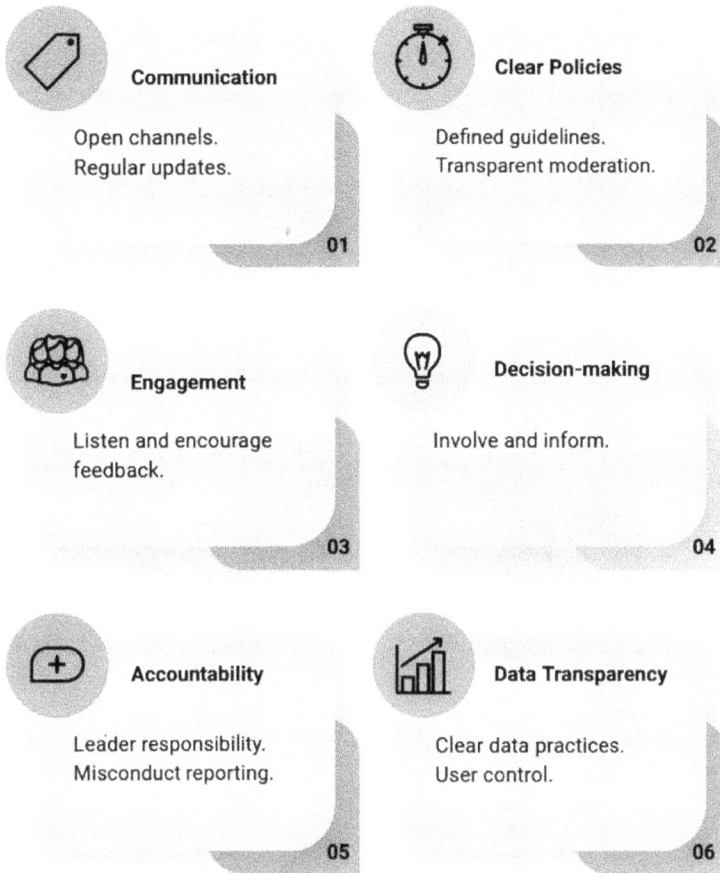

Communication

Open channels.
Regular updates.

01

Clear Policies

Defined guidelines.
Transparent moderation.

02

Engagement

Listen and encourage
feedback.

03

Decision-making

Involve and inform.

04

Accountability

Leader responsibility.
Misconduct reporting.

05

Data Transparency

Clear data practices.
User control.

06

FIGURE 9.1   Transparency in community governance

9.1.3   HANDLING USER CONFLICT AND DISPUTES

User disagreements and disputes increase as these platforms gain popularity (Kiesler, 2014). These confrontations might result from disagreements, misunderstandings, or community rules. Managing these disputes ethically and successfully is essential for a healthy and inclusive online community. User disputes in online communities are resolved using various conflict management tactics. Online community conflict management entails preventing disputes through explicit guidelines and an accepting environment, resolving conflicts by recognizing problems and coming up with reasonable answers, controlling the escalation process through moderator intervention, and transforming relationships via dialogue and understanding (Figure 9.2).

Online communities use active listening, empathy, respect, objectivity, and other strategies as part of their conflict management practices (Figure 9.3). Asking clarifying questions to ensure understanding and paying close attention to both sides are all components of active listening. By sharing emotions, empathy promotes understanding and trust. Respect fosters positive dialogue by recognizing the value of each individual. Neutrality is preserved for just resolutions through objectivity.

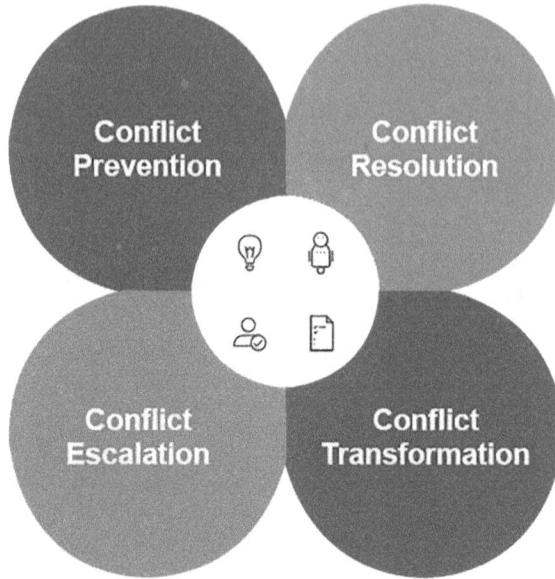

FIGURE 9.2   Factors affecting online community conflict resolution

FIGURE 9.3   Techniques for handling conflicts effectively in online communities

9.1.4   BALANCING FREEDOM OF EXPRESSION WITH RESPONSIBLE CONDUCT

Promoting free speech and user responsibility in online communities requires a delicate balance. While encouraging free speech, boundaries must be established to prevent detrimental conduct. Infractions of community regulations ought to proscribe hate speech, harassment, and disinformation. In order to preserve this equilibrium, community administrators must diligently oversee interactions and promptly rectify any infractions. In addition to promoting a harmonious and diverse environment, they should address abuse and toxicity expeditiously. A moderate amount of transparency can foster community confidence and demonstrate equity. Online communities can foster responsible conduct and protect free speech by encouraging constructive discourse and respectful dialogue. Promoting empathy and respect by encouraging users to consider the impact of their words and actions on others enhances the online ecosystem.

9.2 PRIVACY AND SECURITY CONCERNS IN SOCIAL NETWORKING PLATFORMS

Privacy and security concerns may greatly impact users' willingness to share information on social networking sites and how much trust they place in those platforms. Users who worry about their security and privacy are less likely to join online communities, engage in online interactions, and divulge personal information [9]. Furthermore, a lack of trust in social media sites may affect how they are used and accepted, eventually decreasing their popularity and efficacy.

Numerous strategies have been implemented to address privacy and security issues on social networking sites. These solutions include authentication protocols, security and privacy setups, and internal security measures [10, 11]. To help ensure that access to the website is limited to authorized users only, authentication technologies, photos-of-friend identification, and multi-factor authentication can be used. Users can take control of their data and stop unwanted parties from accessing it by choosing security and privacy settings. This goal can be accomplished by implementing internal security mechanisms, including spam filters, false profile detection, and user reporting systems. These methods help protect users from online dangers and preserve the platform's integrity.

9.2.1 USER INFORMATION FROM BREACHES

Data security has grown critical due to the rise in both frequency and sophistication of cyberattacks. In addition to jeopardizing users' personal information, a breach erodes users' confidence in the platform.

Social networking sites need to use strong security measures, such as access controls, encryption methods, and frequent security audits, to reduce the chance of breaches. In order to improve data privacy, platforms should also abide by industry best practices and compliance guidelines like the California Consumer Privacy Act (CCPA) and General Data Privacy Regulation (GDPR).

Platforms are required to promptly and transparently notify affected users and regulatory authorities in the event of a breach. Retaining people's trust and proving accountability requires transparency. In addition, platforms must offer impacted consumers tools and assistance to lessen the possible damage brought about by the hack, such as credit monitoring or identity theft protection services. A proactive strategy that includes ongoing monitoring, obtaining threat intelligence, and educating staff members on cybersecurity best practices is necessary to prevent intrusions.

9.2.2   Providing Opt-In/Opt-Out Features for Data Sharing

Personalized content, targeted advertisements, and enhanced user experience are just a few of the many uses for the massive amounts of user data collected by social media platforms. Nevertheless, platforms must honor users' wishes about data sharing and respect their privacy settings.

Users are given the option to opt in or opt out, allowing them to vote in or out of certain uses and disclosures of their data. Users can actively consent to data sharing through the "opt-in" process, or they can deny participation by using the "opt-out" option. For robust opt-in/opt-out capabilities, Table 9.1 lists the importance, implementation techniques, user experience impact, legal/regulatory considerations, and technical challenges of each significant aspect.

TABLE 9.1

Considerations crucial to the implementation of robust opt-in/opt-out features

Consideration	Implementation Methods	User Experience Impact
Transparency	Clear language in privacy policies, interactive tutorials	Builds trust, reduces confusion
Granularity	Dropdown menus for data categories, customizable sharing settings	Empowers users, may overwhelm some
Accessibility	Prominent placement in user interface, simplified design	Enhances usability, reduces frustration
Consent Mechanisms	Clear opt-in checkboxes, granular consent options	Builds trust, reinforces user control
Revocability	One-click opt-out buttons, intuitive revocation process	Maintains user trust, encourages engagement

9.2.3   Implementing Secure Authentication Measures

Social networking platforms must implement robust authentication mechanisms to safeguard potential threats and prevent unauthorized parties from accessing user accounts and personal information. This can be accomplished through numerous methods, such as biometric authentication, multi-factor authentication (MFA), and rigorous password policies [10, 12].

Before being granted access to their accounts, users must adhere to the MFA security method by presenting a minimum of two unique forms of identification. This method substantially reduces the likelihood of unauthorized access, identity theft, and account breaches. On the contrary, biometric authentication relies on distinctive physical or behavioral attributes, including but not limited to fingerprints, facial recognition, and voice recognition, to authenticate the identity of a user. Additionally, robust password policies are necessary to guarantee secure authentication. This requires using complex passwords comprising lowercase and capital letters, numerals, and special characters. Additionally, it is critical to refrain from employing widely used passwords or easily surmised information and to update these passwords regularly.

Social networking platforms allow users to safeguard their confidential information by offering the option to integrate security and privacy settings alongside these pre-existing safety measures. This involves enabling security measures such as CAPTCHA to prevent spam and other security concerns and configuring permissions to regulate access to private information, including posts, photographs, and other specifics. These particulars may consist of images, posts, and other forms of content. Promoting user awareness regarding the significance of secure authentication methods is a fundamental measure that can be taken to safeguard users' personal information and foster more secure online interactions. Educating users about the potential hazards of accepting friend requests from unfamiliar individuals, clicking on suspicious-appearing links, and divulging private information on social networking sites is imperative. Social networking platforms can establish more robust security and privacy configurations by implementing protocols and tools, including mechanisms for reporting inappropriate content and safeguarding minors against harassment.

9.3　ETHICAL IMPLICATIONS OF ONLINE INFLUENCER CULTURE

New ethical issues have emerged due to the spread of the online influencer culture, and they should be carefully considered. The possibility of lying is one of the most important ethical issues, particularly when distinguishing between real and virtual effects. The reality that individuals make and maintain virtual influences like Shudu Gram has prompted questions about ontology and moral responsibility. Despite their ability to garner a sizable fan base and influence behavior online, these influencers have succeeded in raising awareness of these issues. Virtual influencers raise moral questions about how to distribute moral responsibility while maintaining operational transparency [13].

Social media influencers' influence on cultural narratives and the portrayal of historically underprivileged and underrepresented perspectives raises further ethical concerns. Although studies have demonstrated that influencer marketing can modify attitudes and promote a more welcoming digital space, it can also strengthen current power structures and maintain unfavorable stories. Because of the complexity of the market and the cultural implications of influencer marketing, it is crucial to evaluate the influencer marketing sector via an interdisciplinary and socio-cultural lens [14].

It is crucial to consider the ethical culture and information privacy concerns common among social media users when analyzing the impact of social media influencers. The idea is that there is a favorable correlation between social media users' information-sharing practices and ethical culture. Users are more likely to divulge information when they believe the platform to be ethical [15]. Given this, social media influencers and platforms must uphold a strong ethical position to encourage user trust and protect personal information.

9.3.1　DISCLOSURE OF SPONSORSHIPS AND PARTNERSHIPS

Influencers frequently work with companies and brands to market goods and services to their audience. However, if you do not tell your followers about these partnerships, they can start to doubt you. Therefore, celebs should not hide the fact that they have sponsorships and relationships; transparency is key.

When an influencer has a financial interest in, or a tangible relationship to, the advertised goods or service, they are legally required to disclose this fact. Whether through hashtags like #ad or #sponsored, vocal acknowledgments in videos, or captions in social media posts, this disclosure must be clear and easy for the audience to see.

Influencers maintain the credibility and genuineness of their material by being open about their partnerships and endorsements. In addition, it lets followers determine the legitimacy and intentions of the promoted content with more knowledge. In order to safeguard consumers from misleading marketing strategies and promote ethical advertising practices, regulatory organizations and advertising standards authorities frequently demand such disclosures.

9.3.2 Authenticity and Transparency in Content Creation

Whereas transparency is primarily concerned with being open with all facts that can impact the pertinent stakeholders of the company, authenticity is more closely linked to image consistency and perceived genuineness [16]. When it comes to content creation, transparency is removing superfluous layers from the story and presenting it in its most unadulterated form. It also entails disclosing the creative process, sharing real stories, and admitting mistakes. The audience connects with this genuineness on a more profound level.

Authenticity and transparency go hand in hand. Modern consumers value integrity and honesty. Not only does sharing a brand's true values, objectives, and struggles humanize the story, but it also creates a more genuine and approachable image for the company. Content transparency accomplishes various goals, such as building consumer loyalty and improving brand reputation. Brands can optimize their content strategies by investigating these goals.

Transparency in User Experience (UX) refers to a more comprehensible and transparent experience. When items or services are presented truthfully, users find it easier to make decisions. Transparency in UX is an emotive component that fosters user trust in addition to being informational.

9.3.3 Avoiding Deceptive Marketing Practices

To avoid misleading advertising, influencers and brands should disclose all substantial links to relevant brands and goods in posts, stories, and online content. Material connections include payments, commissions, free, loaned, or reduced goods, services, travel, tickets, or personal or family relationships. Disclosures should be apparent, not in fine print or links.

Influencers should provide honest evaluations, comments, and testimonies based on personal experience. Brands should constantly verify that influencers are only making performance claims on their behalf if they are company-endorsed and based on acceptable testing.

Brands should also avoid "astroturfing," which involves funding bogus consumer reviews or deception. Companies should constantly inform influencers of their policies and expectations and create or maintain influencer marketing best practices.

International regulators, including the Canadian Competition Bureau and the Advertising Standards Council of India, have proposed rules to regulate influencer marketing and prevent deception. The Indian Consumer Protection Act 2019 defines a deceptive advertisement as one that misrepresents a product or service's nature, quantity, or quality. The Act punishes unfair trading practices that misrepresent goods and services "quality, quantity, grade, composition, style, or model."

9.3.4 IMPACT OF INFLUENCER BEHAVIOR ON AUDIENCE

The behavior of influencers can exert a significant influence on their audience. This influence is often multifaceted and can have both positive and negative ramifications. Understanding the impact of influencer behavior on the audience requires a nuanced analysis of several key factors.

Audience Perception Analysis. Employing sentiment analysis and natural language processing (NLP) techniques to gauge the audience's perception of influencer behavior. This involves monitoring social media conversations, comments, and reactions to assess the overall sentiment toward the influencer's actions.

Behavioral Modeling. Utilizing machine learning algorithms to model and predict how specific influencer behaviors may influence audience actions and attitudes. This involves analyzing historical audience engagement and reaction data to identify patterns and correlations between influencer actions and audience responses.

Ethical Framework Evaluation. Applying ethical frameworks such as consequentialism, deontology, and virtue ethics to evaluate the ethical implications of influencer behavior on the audience. This involves considering factors such as honesty, transparency, and promoting positive values in influencer content.

Psychological Impact Assessment. Conducting psychological studies and surveys to assess the psychological impact of influencer behavior on audience members. This may involve measuring variables such as self-esteem, body image perception, and purchasing behavior influenced by exposure to influencer content.

Long-Term Effects Analysis. Examining the long-term effects of consistent exposure to influencer content on audience behavior and attitudes. This involves longitudinal studies and data analysis to track changes in audience perceptions and behaviors over time.

Cultural Context Consideration. Taking into account cultural differences and norms when assessing the impact of influencer behavior on diverse audience demographics. This involves conducting cross-cultural studies and sensitivity analyses to ensure findings are applicable across different cultural contexts.

9.4 CYBERBULLYING AND TOXIC BEHAVIOR IN SOCIAL MEDIA

Cyberbullying is the practice of harassing, threatening, or bullying someone online, usually in secret, using social networking sites and electronic gadgets [17, 18]. It might manifest in various ways, such as distancing oneself from online communities, distributing rumors, posing as someone else, or exchanging furious or threatening messages. Because offensive material may spread quickly and be permanently stored online, cyberbullying is especially dangerous. Cyberbullying victims are more likely to experience anxiety, despair, substance misuse, food and sleep issues, as well as poorer academic outcomes [17]. Suicidal thoughts and self-injurious behavior have also been connected to cyberbullying [19, 20]. Cyberbullying might be more detrimental than traditional in-person bullying due to factors, including anonymity and the ease with which harmful content can be quickly disseminated online [17].

Social media companies have introduced policies and technologies targeted at reducing online abuse and cyberbullying in order to solve this problem. However, continued observation and investigation are necessary to ensure the efficacy of these initiatives [17]. Encouraging empathy, digital literacy, and responsible online conduct in individuals is also critical. To effectively combat cyberbullying and build safer online settings, a multidimensional approach incorporating technology, education, and community support is ultimately required [17, 20, 21].

9.4.1 STRONG ANTI-BULLYING POLICIES

Schools must define bullying, analyze its causes, build on current policies, and involve the full school community to design successful anti-bullying measures. Schools must acknowledge that bullying includes verbal, social, and cyberbullying, power imbalances, and recurrent actions.

Anti-bullying measures should target systemic biases, school culture, and environmental variables, as well as individual behaviors. Schools may improve their anti-bullying policies and foster a culture of respect, safety, and inclusivity by fostering open communication between students, teachers, parents, and administrators.

9.4.2 ENCOURAGING REPORTING AND SUPPORT SYSTEMS

Cyberbullying occurrences can be better identified and addressed with the use of reporting systems. Victims can also benefit from support networks, which offer assistance and encourage positive conduct.

When it comes to combating cyberbullying, fostering an environment of openness and honesty is key. To achieve this goal, it is necessary to create transparent channels for reporting crimes, aid victims, and hold offenders accountable. An example of this is the English National Health Service, where research on patient safety indicated that the use of excellent administrative processes may achieve openness, safety, effectiveness, sensitivity to patient needs, and ongoing inquiry, learning, and improvement [22].

Technology companies are responsible for ensuring that young people may safely use the internet. Included in this is the development of platforms and tools that encourage constructive interactions, the sharing of data pertinent to the health effects of platforms with independent researchers, and the establishment of efficient and prompt mechanisms to resolve complaints and requests from youth, families, and educators regarding harmful content and online abuse.

9.4.3 COMBATING TROLLING AND HARASSMENT

The goal of trolling, a type of cyberbullying, is to get people to react negatively. Therefore, the perpetrators willfully engage in disruptive or provocative behavior. The effects on its targets can be devastating, manifesting in various ways, including flaming, trolling asymmetry, and hate speech [23]. On the flip side, harassing behavior is characterized by persistent and unwanted actions that are meant to upset or hurt the target.

Having well-defined rules and regulations for conduct on the internet is crucial in the fight against trolling and harassment. Trolling and harassment should be defined in these policies, and the repercussions for breaking them should also be outlined. The significance of recording harassment occurrences, notifying the relevant response teams, and offering victim support services is highlighted in this handbook.

In the fight against harassment and trolling, tech businesses can be just as important as policymakers. There is a need for more efficient content triage and proactive moderation, and artificial intelligence (AI) and machine learning algorithms can help with that. For example, in cases when policy violations were discovered and addressed before public disclosure, major platforms like Google, Facebook/Meta, and Twitter have made the proportion public. Scholars, lawmakers, and governments have all demanded increased oversight of internet platforms' actions, but the moderation process still needs more transparency and accountability.

9.4.4   Providing Mental Health Resources for Users

The potential of self-management of e-resources to empower individuals to participate in their care and improve outcomes actively has been emphasized in research as an important aspect of mental health care. To make the most of these resources, we need defined standards for their creation and reporting since the evidence base is still growing [24].

It is widely acknowledged that including users in mental health services can greatly improve the quality and outcomes of care. Participation from service users in the design and assessment of mental health services has been associated with better, more tailored treatment by capitalizing on service recipients' distinct backgrounds and viewpoints. The strategy takes a more holistic and user-centered approach to mental health assistance, improving service quality while promoting social inclusion, empowerment, and therapeutic benefits for users.

New digital mental health tools show great promise for enhancing mental healthcare delivery, particularly to youth. Potentially alleviating the impact of mental disease and increasing access to treatment for vulnerable populations, such as children and adolescents, these technologies present chances to track, detect, and handle mental health demands in an early stage. Deploying digital mental health solutions for young users requires careful attention to ethical factors, including patient autonomy, equality of access, and privacy [25].

9.5   ETHICAL USE OF DATA AND USER ANALYTICS IN SOCIAL NETWORKING

Privacy, permission, and the digital divide are ethical questions from social media analytics and data collection. Data collected from users by social media platforms can be utilized for various purposes, such as research and marketing [26]. As the *Cambridge Analytica* incident has shown, however, privacy breaches can result from acquiring and using such data [26]. In this case, a friend had given the firm access to the user's personal information without the user's knowledge or consent.

There is also the ethical problem of the digital divide, which arises when policies are based on data acquired through sensors, social media, etc., and these policies may only consider the requirements of those who have access to these modern means, which could leave certain persons or parts of society out. There may be privacy and security concerns because consumers are less likely to know what data major web service providers, like Google and Facebook, collect from them when they utilize Single Sign On services.

Simple joins between these anonymous datasets and other datasets can de-anonymize them in a relatively short amount of time; therefore, more than common anonymization strategies, such as substituting a user's personal information in a dataset, may be required to ensure privacy. Instead of proposing more complex anonymization methods, a stronger data access policy should be implemented to tackle these ethical concerns.

9.5.1 OBTAINING INFORMED CONSENT FOR DATA COLLECTION

To get research subjects' informed permission, researchers must explain the study's purpose, any hazards or advantages, and their voluntary participation rights. Presenting information in a way that is easy to understand and comprehend is crucial to avoid undue influence or coercion. To maintain the ethical standards of research involving personal data, the informed consent process should contain specifics concerning data protection, confidentiality, privacy safeguards, and how any incidental discoveries would be handled [27].

Alternative models or waivers of informed consent may be considered subject to stringent requirements when acquiring it proves difficult or impracticable, such as research involving vulnerable populations or emergency care. Nevertheless, it is crucial to prioritize ethical issues and make every effort to respect the autonomy and rights of participants, especially in situations when consent is waived, to ensure that the research process remains intact [28, 29]. When working with participants who come from varied backgrounds, have low reading levels, or have less autonomy, researchers have additional challenges in navigating the informed consent process and improving comprehension through communication.

Researchers and institutions need to ensure participants understand everything going on during the permission process to overcome the difficulties and hurdles of getting informed consent for data collecting. The ideals of respect for individuals' autonomy and rights in research settings can be upheld by researchers who adhere to ethical rules, give participants enough time to think about their decisions, and provide information clearly [27].

9.5.2 DATA ANONYMIZATION AND AGGREGATION

Data anonymization methods include masking, pseudonymization, aggregation, randomness, generalization, and swapping. These methods change sensitive data to comply with privacy rules and allow its use. Aggregating data from several sources creates a larger dataset for data analysis or software testing. Since it is harder to identify individual users, aggregated data can provide significant insights while reducing the risk of re-identification.

To prevent re-identification and comply with data privacy rules, anonymization masks personally identifiable (PII) data, such as names, addresses, and other sensitive data. This method keeps data functional and protects user privacy. Businesses must anonymize data to comply with privacy laws like the GDPR and CCPA. These restrictions require corporations to preserve customer data and allow anonymity. Data anonymization helps firms comply with these rules and maintain client trust.

9.5.3 Data for Improving User Experience Responsibly

Using data to improve user experiences demands a difficult balance between innovation and ethics. By monitoring user interactions, preferences, and behaviors, companies can adjust their products and services to varied user needs [30]. This data-driven strategy lets organizations personalize experiences, optimize user interfaces, and boost satisfaction.

Responsible data usage improves user experience by gathering, evaluating, protecting user privacy, and being transparent. Organizations must follow data protection laws like the GDPR to protect user data and individual rights [31, 32]. Data governance must include minimization, encryption, and secure storage to reduce data breaches and unauthorized access.

The ethical implications of data used for user experience improvement include justice, prejudice prevention, and inclusion. Organizations must actively address data collecting and algorithmic decision-making biases to prevent discrimination and provide equal user experiences [30]. Companies may improve product and service inclusion and fairness by including various perspectives, bias audits, and fairness-aware algorithms.

9.5.4 Avoiding Discriminatory Practices in Data Analysis

Several approaches can be used to circumvent biased data analysis. Differential impact, redlining, statistical discrimination, differential treatment, and direct discrimination must first be defined and understood [33]. The use of these definitions aids in the detection and remediation of discriminating data analytics results.

People who are already disadvantaged may be among the targets of data mining tools. Data mining algorithms must be crafted to prevent exploitation and minimize harm to these populations. Making a clear differentiation between neutral or fair discrimination and damaging and unfair discrimination is crucial [33].

Evaluating and auditing datasets and algorithms for potentially biased impacts is required to prevent discriminatory outcomes in data analytics. Data gathering, machine learning procedures, training materials, and category definitions are all potential sources of bias that reviewers should consider.

Following data privacy legislation, including the Equality Act and the GDPR, is important. These statutes safeguard particular populations from unjust treatment and mandate nondiscrimination. It is important to weigh the potential risks to other rights and freedoms of individuals with the potential benefits of collecting more data about statistically small minority groups to equalize the proportions of correct predictions, especially if a relative lack of data about those groups drives the model's discriminatory outcomes.

9.6 PROMOTING DIGITAL CITIZENSHIP AND RESPONSIBLE ONLINE ENGAGEMENT

9.6.1 USERS ON DIGITAL LITERACY AND CRITICAL THINKING

The capacity to find, assess, use, and generate information through the use of digital technology, communication tools, or networks is known as digital literacy. Critical thinking is the capacity to examine and assess data, concepts, and arguments in order to arrive at well-informed conclusions.

Both digital literacy and the ability to think critically are supplementary to one another. Critical thinking equips people to assess the reliability, accuracy, and applicability of information they find online, while digital literacy allows them to access and utilize that data efficiently. People who can think critically are better able to communicate and work together in virtual spaces and spot and avoid cyberbullying, disinformation, and manipulation.

Using technology effectively and thinking critically are prerequisites for engaging in online politics, culture, and economics. Responsible and ethical internet use for political, economic, social, and cultural purposes is crucial to digital citizenship. Individuals can better understand and use their rights and responsibilities as digital citizens when they are literate and think critically about the digital world.

9.6.2 RESPECTFUL COMMUNICATION AND INTERACTION

Digital citizenship requires respectful communication and interaction. Effective communication requires suitable language and nonverbal indicators like body language and tone of voice to ensure accurate and courteous reception. In the digital world, polite communication and interaction are crucial because online environments can be anonymous or impersonal, reducing accountability [34].

Communication between patients and healthcare providers is crucial for high-quality care. Nurses and midwives communicating properly and showing empathy and respect make patients happier. Poor communication, such as verbal abuse, disdain, or denial of inquiry, can negatively impact patients' views of services. Patients and care providers should speak and be heard without interrupting, ask questions for clarity, voice their ideas, and share information [35].

Building and maintaining great online relationships requires respectful communication and involvement. Digital citizenship means utilizing technology ethically and respecting others' privacy. Avoid cyberbullying, internet harassment, and other rude actions. Digital citizenship also requires understanding and accepting the repercussions of online behavior. Respectful online speech and interaction can make the internet safer and more inclusive.

9.6.3 BUILDING EMPATHY AND UNDERSTANDING IN ONLINE COMMUNITIES

Empathy-building in online courses has been demonstrated through research to be a useful strategy for encouraging good interactions and fostering a supportive learning environment. Eight similar themes of practice are used by faculty to foster empathy in the context of online higher education, according to a study that looked at the empathic practices of successful higher education teachers from seven universities that offer online programs. Some of these themes are developing relationships, fostering a supportive learning environment, encouraging active participation, utilizing compassionate communication techniques, giving tailored feedback, fostering peer interaction, utilizing real-world examples, and setting an example of compassionate behavior.

Outside the forum, experiences with the beginning and diagnosis of illness create emotional and informational requirements that are the first steps toward developing empathy. The forum's friendly culture and sympathetic tone are established by users discovering that others share and understand their needs and experiences through connections made in this compassionate environment based on resemblance, bonds, and shared emotions, as well as empathy functions [35].

In addition to establishing a healthy learning environment, developing empathy and understanding in online communities is essential for encouraging pleasant social interactions and advancing community development. Empathy is fundamental to local growth at the community level and is the cornerstone of community and local development. A community is a sociological and psychological construct that embodies a location, its inhabitants, and the connections that exist within it. Social interactions within the community serve as a foundation for intentional group initiatives to enhance social well-being. These exchanges raise people's awareness of regional concerns and issues and may inspire the creation of targeted initiatives to address community needs [36].

9.6.4 Users to Contribute Positively to Online Discourse

To help their students identify biases, determine the trustworthiness of sources, and differentiate between trustworthy and untrustworthy material, educators should teach them the art of critical thinking and evaluation regarding internet content. Students must adopt this critical thinking style to participate in online debates with respect and knowledge.

The education of appropriate online conduct, emphasizing privacy issues, cyberbullying awareness, and ethical considerations, is another component of digital literacy. Teachers can assist their pupils in being more self-aware of the digital footprint they leave behind by stressing the necessity of being respectful of intellectual property and the repercussions of their online activities.

The next generation of digital natives will be better prepared to handle the challenges of today's interconnected world, thanks to digital literacy programs that teach children to think critically and behave responsibly while using the internet. Active engagement and the application of digital skills can be fostered through collaborative projects, multimedia assignments, and interactive simulations. Participating in coding activities has multiple benefits, including improving problem-solving abilities and better grasping how digital systems work.

Educators may play a key role in fostering digital citizenship by highlighting the need for positive online behaviors, empathy, and responsible resource use. Promoting constructive online conversation requires the establishment of a school culture that upholds the principles of ethical digital behavior and courteous online communication. Workshops and instructional programs that encourage parental participation in digital literacy initiatives provide parents with the information they need to help their children use technology responsibly at home.

REFERENCES

1. Ferreira, J.J., Fernandes, C., Veiga, P.M. and Rammal, H.G., Ethics and the dark side of online communities: Mapping the field and a research agenda. *Information Systems and e-Business Management*, 2023: pp. 1–25.
2. Sugiura, L., R. Wiles, and C. Pope, Ethical challenges in online research: Public/private perceptions. *Research Ethics*, 2017. 13(3–4): pp. 184–199.
3. Cilliers, L. and K. Viljoen, A framework of ethical issues to consider when conducting internet-based research. *South African Journal of Information Management*, 2021. 23(1): pp. 1–9.
4. Hokke, S., Hackworth, N.J., Quin, N., Bennetts, S.K., Win, H.Y., Nicholson, J.M., Zion, L., Lucke, J., Keyzer, P. and Crawford, S.B., Ethical issues in using the Internet to engage participants in family and child research: A scoping review. *PloS one*, 2018. 13(9): p. e0204572.
5. Fiesler, C., Zimmer, M., Proferes, N., Gilbert, S. and Jones, N., Remember the human: A systematic review of ethical considerations in reddit research. *Proceedings of the ACM on Human-Computer Interaction*, 2024. 8(GROUP): pp. 1–33.
6. Douglas, S. and A. Meijer, Transparency and public value—Analyzing the transparency practices and value creation of public utilities. *International Journal of Public Administration*, 2016. 39(12): pp. 940–951.
7. Ortega-Rodríguez, C., A. Licerán-Gutiérrez, and A.L. Moreno-Albarracín, Transparency as a key element in accountability in non-profit organizations: A systematic literature review. *Sustainability*, 2020. 12(14): p. 5834.
8. Shields, K.F., Moffa, M., Behnke, N.L., Kelly, E., Klug, T., Lee, K., Cronk, R. and Bartram, J., Community management does not equate to participation: Fostering community participation in rural water supplies. *Journal of Water, Sanitation and Hygiene for Development*, 2021. 11(6): pp. 937–947.
9. Ali, S., Islam, N., Rauf, A., Din, I.U., Guizani, M. and Rodrigues, J.J., Privacy and security issues in online social networks. *Future Internet*, 2018. 10(12): p. 114.
10. Jain, A.K., S.R. Sahoo, and J. Kaubiyal, Online social networks security and privacy: Comprehensive review and analysis. *Complex & Intelligent Systems*, 2021. 7(5): pp. 2157–2177.
11. Mutambik, I., Lee, J., Almuqrin, A., Halboob, W., Omar, T. and Floos, A., User concerns regarding information sharing on social networking sites: The user's perspective in the context of national culture. *Plos One*, 2022. 17(1): p. e0263157.
12. Sagar, K. and V. Waghmare, Measuring the security and reliability of authentication of social networking sites. *Procedia Computer Science*, 2016. 79: pp. 668–674.
13. Kim, D. and Z. Wang, The ethics of virtuality: Navigating the complexities of human-like virtual influencers in the social media marketing realm. *Frontiers in Communication*, 2023. 8: p. 1205610.
14. Gurrieri, L., J. Drenten, and C. Abidin, *Symbiosis or parasitism? A framework for advancing interdisciplinary and socio-cultural perspectives in influencer marketing.* 2023: Taylor & Francis. pp. 911–932.
15. Chai, S., Does cultural difference matter on social media? An examination of the ethical culture and information privacy concerns. *Sustainability*, 2020. 12(19): p. 8286.

16. Coker, K., Howie, K., Syrdal, H., Vanmeter, R. and Woodroof, P., The truth about transparency and authenticity on social media: How brands communicate and how customers respond: An abstract, in *Back to the Future: Using Marketing Basics to Provide Customer Value: Proceedings of the 2017 Academy of Marketing Science (AMS) Annual Conference*. 2018, Springer.

17. Abaido, G.M., Cyberbullying on social media platforms among university students in the United Arab Emirates. *International Journal of Adolescence and Youth*, 2020. 25(1): pp. 407–420.

18. Kee, D.M.H., M.A.L. Al-Anesi, and S.A.L. Al-Anesi, Cyberbullying on social media under the influence of COVID-19. *Global Business and Organizational Excellence*, 2022. 41(6): pp. 11–22.

19. Schodt, K.B., Quiroz, S.I., Wheeler, B., Hall, D.L. and Silva, Y.N., *Cyberbullying and mental health in adults: The moderating role of social media use and gender. Frontiers in Psychiatry*, 2021. 12: p. 674298.

20. Garett, R., L.R. Lord, and S.D. Young, Associations between social media and cyberbullying: A review of the literature. *Mhealth*, 2016. 2: pp. 1–7.

21. Chatzakou, D., Leontiadis, I., Blackburn, J., Cristofaro, E.D., Stringhini, G., Vakali, A. and Kourtellis, N., Detecting cyberbullying and cyberaggression in social media. *ACM Transactions on the Web (TWEB)*, 2019. 13(3): pp. 1–51.

22. Martin, G., Chew, S., McCarthy, I., Dawson, J. and Dixon-Woods, M., Encouraging openness in health care: Policy and practice implications of a mixed-methods study in the English National Health Service. *Journal of Health Services Research & Policy*, 2023. 28(1): pp. 14–24.

23. Ortiz, S.M., Trolling as a collective form of harassment: An inductive study of how online users understand trolling. *Social Media+ Society,* 2020. 6(2): p. 2056305120928512.

24. Karasouli, E. and A. Adams, Assessing the evidence for e-resources for mental health self-management: A systematic literature review. *JMIR Mental Health*, 2014. 1(1): p. e3708.

25. Wies, B., C. Landers, and M. Ienca, Digital mental health for young people: A scoping review of ethical promises and challenges. *Frontiers in Digital Health*, 2021. 3: p. 697072.

26. Zhu, X., Q. Cao, and C. Liu, Mechanism of platform interaction on social media users' intention to disclose privacy: A case study of Tiktok app. *Information*, 2022. 13(10): p. 461.

27. Kadam, R.A., Informed consent process: A step further towards making it meaningful! *Perspectives in Clinical Research*, 2017. 8(3): pp. 107–112.

28. Laurijssen, S.J., van der Graaf, R., van Dijk, W.B., Schuit, E., Groenwold, R.H., Grobbee, D.E. and de Vries, M.C., When is it impractical to ask informed consent? A systematic review. *Clinical Trials*, 2022. 19(5): pp. 545–560.

29. O'Sullivan, L., Feeney, L., Crowley, R.K., Sukumar, P., McAuliffe, E. and Doran, P., An evaluation of the process of informed consent: Views from research participants and staff. *Trials*, 2021. 22: pp. 1–15.

30. Mikalef, P., Conboy, K., Lundström, J.E. and Popovič, A., *Thinking responsibly about responsible AI and 'the dark side'of AI*. 2022: Taylor & Francis. pp. 257–268.

31. Balayn, A., C. Lofi, and G.-J. Houben, Managing bias and unfairness in data for decision support: A survey of machine learning and data engineering approaches to identify and mitigate bias and unfairness within data management and analytics systems. *The VLDB Journal*, 2021. 30(5): pp. 739–768.

32. Zwitter, A. and O.J. Gstrein, *Big data, privacy and COVID-19–learning from humanitarian expertise in data protection*. 2020: Springer. pp. 1–7.

33. Favaretto, M., E. De Clercq, and B.S. Elger, Big Data and discrimination: Perils, promises and solutions. A systematic review. *Journal of Big Data*, 2019. 6(1): pp. 1–27.

34. Willett, J.F., LaGree, D., Shin, H., Houston, J.B. and Duffy, M., The role of leader communication in fostering respectful workplace culture and increasing employee engagement and well-being. *International Journal of Business Communication*, 2023: p. 23294884231195614.

35. Kwame, A. and P.M. Petrucka, A literature-based study of patient-centered care and communication in nurse-patient interactions: Barriers, facilitators, and the way forward. *BMC Nursing*, 2021. 20(1): p. 158.

36. Berardi, M.K., White, A.M., Winters, D., Thorn, K., Brennan, M. and Dolan, P., Rebuilding communities with empathy. *Local Development & Society*, 2020. 1(1): pp. 57–67.

10 Ethics of IT Governance and Policy

10.1 ETHICAL CONSIDERATIONS IN IT GOVERNANCE FRAMEWORKS

Effectiveness and efficiency should be promoted by a strong governance structure, which should also uphold moral principles that correspond with society's goals. This requires the formulation of clear standards and broad guidelines that include ethical issues at each phase of the decision-making process. Information technology governance needs ethical decision frameworks as a key component. These frameworks offer systematic ways to investigate the moral consequences of various deeds and choices. These frameworks often guide stakeholders when discussing difficult ethical issues, such as utilitarianism, deontology, and virtue ethics. Companies that apply a systematic framework for carrying out ethical analysis can ensure that their information technology governance processes are based on moral reasoning.

Different ethical dilemmas that need to be thoroughly examined and discussed are constantly present in information technology administration. Applying case studies and ethical scenarios can clarify these challenges, which will encourage stakeholders to carefully consider the ethical consequences of the decisions they make in the matter. Businesses can create plans to deal with the moral dilemmas that emerge in the management of information technology, and by assessing real-world scenarios, they can also learn more about these problems.

Businesses that implement systems for ethical evaluation and auditing improve their capacity to spot and correct ethical transgressions before they become more serious issues. These systems are crucial tools that play a major role in preserving responsibility and openness inside the limits of information technology governance. All those participating in the monitoring process should be encouraged to have an ethical consciousness and sense of responsibility. This is a necessary part of moral oversight. This underlines the need for moral conduct in all facets of the management of information technology duties.

10.1.1 Incorporating Ethical Principles into IT Governance Structures

If companies are to include moral concepts into their IT governance frameworks successfully, they must negotiate the complicated realm of artificial intelligence ethics and governance. A growing body of academic research emphasizes the need for artificial intelligence governance to ensure that business activities comply with moral principles. This means that the larger organizational governance structure has to include corporate governance, data governance, information technology governance, and AI governance. Academics stress the need to formalize artificial intelligence (AI) governance to support accountability frameworks and handle ethical concerns at many levels, from regulatory oversight to AI developers. More studies are needed to ascertain how companies should manage their AI systems [1]. Despite the emphasis on AI ethics, definitions of AI governance still need to be clarified.

The situation now demonstrates how challenging it is to uphold moral standards about artificial intelligence because frameworks sometimes do not offer sufficient penalties for transgression. How well AI ethics frameworks will direct the development of ethical AI has been questioned because these frameworks need enforcement mechanisms. The focus on ethics and the absence of strong regulations enable technology businesses to function without constraints, protecting them from possible legal repercussions. This emphasizes how urgently stronger governance structures that can efficiently manage artificial intelligence technology and ensure the preservation of moral principles are needed [2].

National committees and bodies entrusted with monitoring the global advancement of artificial intelligence support the adoption of ethics-oriented governance of the technology [3]. These groups are essential to ensuring that artificial intelligence is used morally and advancing long-term adaptable governance solutions. The need for regulating agencies supervising the creation of artificial intelligence, carrying out effect assessments, and ensuring morality by certifying and testing algorithms is emphasized. Businesses can successfully manage the problems related to artificial intelligence governance and work toward creating AI systems that benefit society if they operationalize ethical and human rights concepts [3].

Considering the current discussion on the ethics and governance of artificial intelligence, businesses are strongly advised to develop a comprehensive plan for AI governance that incorporates moral principles into the frameworks that control information technology. The strategy being followed entails matching the governance of artificial intelligence with corporate objectives and ensuring that ethical issues are ingrained in the actions of individual designers as well as the culture of the company. Companies that adopt normative ethical governance as a cornerstone of responsible research and innovation can negotiate the ethical issues of creating artificial intelligence and promote an ethical culture within the AI ecosystem [4].

10.1.2 Ethical Decision Frameworks for IT Governance

In information technology administration, moral decisions need to be made under difficult conditions where social and moral issues coexist with technical developments. Many models and frameworks have been created to help companies make ethically right judgments. These frameworks offer systematic ways to analyze problems, assess stakeholder interests, and identify possible workable solutions. Table 10.1 compares IT governance ethical decision-making models and their focus areas:

As Figure 10.1 illustrates, a framework based on principles for IT governance should stress important moral precepts such as beneficence, autonomy, fairness, and non-maleficence.

TABLE 10.1
IT governance ethics model comparison

Ethical Decision-Making Models	Consequences Analysis	Rule Adherence	Character Focus	Rights Protection
Utilitarianism	✓			
Deontological Ethics		✓		
Virtue Ethics			✓	
Rights-Based Approaches				✓

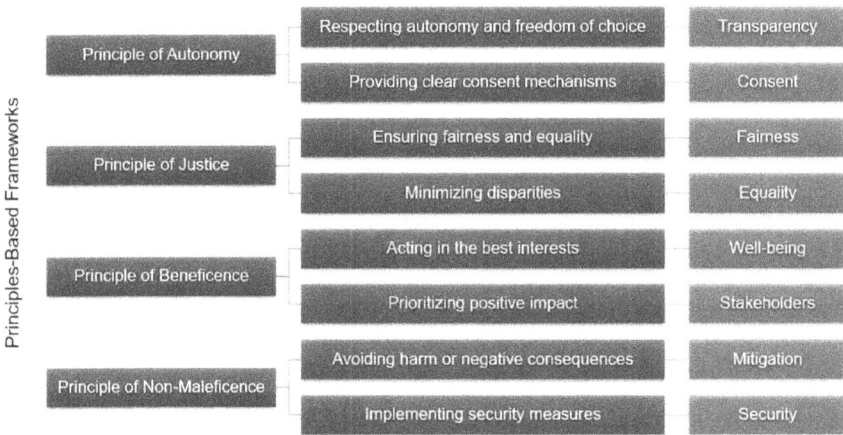

FIGURE 10.1   Systems of IT governance based on principles

10.1.3 ETHICAL DILEMMAS IN IT GOVERNANCE

The ethical conundrums in information technology governance frequently center on technology use transparency, security, accountability, and privacy concerns.

Research ethics issues can be difficult, particularly when combined with corporate and financial ethics. One must give these matters serious thought. This study area has many problems to solve, including those with the moral use of AI, legality, control, and financial harm [5]. IT governance that works must strike a balance between accountability, economic opportunity, and the moral ramifications of artificial intelligence's possible impact on society. Efficient handling of these ethical conundrums requires an understanding of the benefits and risks connected with the application of technology in many fields.

Views of researchers, members of research ethics boards, and industry professionals regarding ethical issues in information technology governance shed light on the variety of these issues in academia. Regarding establishing ethical procedures in research and technology development, research integrity, conflicts of interest, respect for research participants, and ethical discomfort are at the forefront of the discussion [6].

10.1.4 Ethical Oversight Mechanisms in IT Governance

AI governance is crucial to reaping the rewards of artificial intelligence systems and controlling their risks. It should be included in the governance framework of an organization together with data, information technology, and corporate governance. Layered AI governance systems that include ethical and legal levels, as well as levels ranging from AI developers to regulation and oversight, have been proposed by Mäntymäki et al. [1]. The promotion of actual accountability systems is the main goal of the third wave of ethical artificial intelligence research.

A basic issue in the debate about artificial intelligence governance is data governance. This is because of the many organizational and technological obstacles that prevent efficient data control and the determination of responsibility. The information imbalances between regulators and technology businesses mean the current regulatory and governance structures are ill-prepared to handle the societal issues that artificial intelligence (AI) brings. This is so because the information asymmetries make it harder for regulators to close informational gaps and because of the quick progress in technology. In reaction to issues brought about by "hard" regulatory frameworks, industry bodies and governments have been creating self-regulatory or "soft law" methods to regulate artificial intelligence creation more and more; nonetheless, the efficacy of these measures is still restricted [7].

Artificial intelligence research ethics responsibilities are essential to ethical oversight systems in information technology governance. Governments might gain by doing public education and ethical training to increase public knowledge and sensitize people to research ethics in artificial intelligence. Stakeholders cannot be held accountable only for the regulatory and decision-making procedures related to artificial intelligence; these procedures must also be based on legal principles. Previously assigned to specialists or researchers who use technology, these responsibilities will now be attributed to digital mental health apps and other entities. Concerning the possibility that biased algorithms could be supplied to AI models, it is critical that researchers and developers of artificial intelligence exercise extreme vigilance. Clinicians will need to be able to speak with patients diplomatically about the results generated by machine learning (ML) models and be aware of the possibility of bias and mistakes [8].

10.2 ETHICAL DECISION-MAKING IN IT POLICY DEVELOPMENT

As the Association for Computing Machinery (ACM) has recognized the need for moral decision-making, it has updated its formal code of ethics for the first time since 1992 [9]. However, no significant difference between the control group and those who received clear instructions to consider the ACM code of ethics when making decisions [9]. This implies that finding techniques that can improve the process of making ethical decisions in software engineering is a challenge for the research community.

A conceptual framework has been presented by Banks et al. [10] for the 21st-century ethical decision-making process. This framework is located inside the more comprehensive framework of ethical decision-making. This paradigm mandates conformity to commonly recognized moral principles and considers the impact of different socioeconomic and cultural groups' legal, professional, or moral standards. One crucial consideration to remember is that ethical norms may vary over time, so the framework should be flexible enough to accommodate changes like revisions to the code of ethics of the American Psychological Association. Most professionals are assumed to want to act morally and "do the right thing" [10]. Hence, the framework aims to increase understanding of ethical issues.

Laws and executive procedures can help to realize practical ethics in formulating policies by raising and lowering practical obstacles and boosting and diminishing intrinsic incentives. It is feasible for outside obligations to either strengthen or weaken the impact of internal incentives. Nevertheless, laws should be used carefully since they could have unexpected effects and because ethical transactions should consider human feelings. People find it easier to act morally in a more ethical atmosphere, and it lowers the expenses of doing so while raising the expenses of doing immoral acts. Implementing suitable laws and executive processes can help establish such an atmosphere, and ethical audits can help progressively remove obstacles [11].

One way to help people and organizations planning and participating in policy discussions navigate ethical issues is to develop an ethical analysis of policy dialogues. Evidence is supported and encouraged in decision-making processes through policy dialogues. Three core ethical principles of policy discussions are the framework for the analysis: (i) the objectives or aims of the policy discussion, (ii) pertinent procedural values, and (iii) suitable substantive values. Usually, procedural values—which are connected to the decision-making process (the way decisions are made) and substantive values—which are connected to the content (the decisions that are made and the reasons for making them), are used to justify decision-making processes [12].

10.2.1 Ethical Impact Assessment in Policy Development

Forming policies requires the Ethical Impact Assessment (EIA) to be a necessary step. It aims to ensure that, before its execution, interested parties fully consider the ethical ramifications of the policy. A concurrently conducted Privacy Impact Assessment (PIA) and an EIA are complementary tools. It gives stakeholders a structure to examine the situation's ethical ramifications and take the necessary preventative measures [13].

Figure 10.2 shows how the EIA addresses privacy and ethical concerns, guides policy debates, and manages policy risks. EIA is key in creating and implementing new information and communications technologies and apps that exploit personal data. This is so because this technology and apps may raise privacy and ethical concerns. EIA can ensure that the ethical issues are considered and that compensating steps are taken before a deployment. The ethical impact assessment framework (EIA) proposed in [13] could be applied to guarantee that mitigation steps are carried out as necessary and that stakeholders sufficiently investigate ethical implications before deployment.

EIA can help people and organizations navigate ethical issues within the scope of policy debates to come to morally sound and well-founded decisions. In many policy domains, including public health, policy dialogues promote and support evidence in policy-making processes and structured, evidence-informed discussions to achieve each goal. Decision-makers in policy debates can benefit from and be justified in their decisions using EIA to help them navigate ethical dilemmas [12].

An important field where EIA is essential is model-based policymaking, which uses mathematical models to assess the risks of human and animal diseases and how to reduce them. Using model evidence in policy, EIA can help guarantee that ethical risks are considered and communicated and that ethical issues are managed [14].

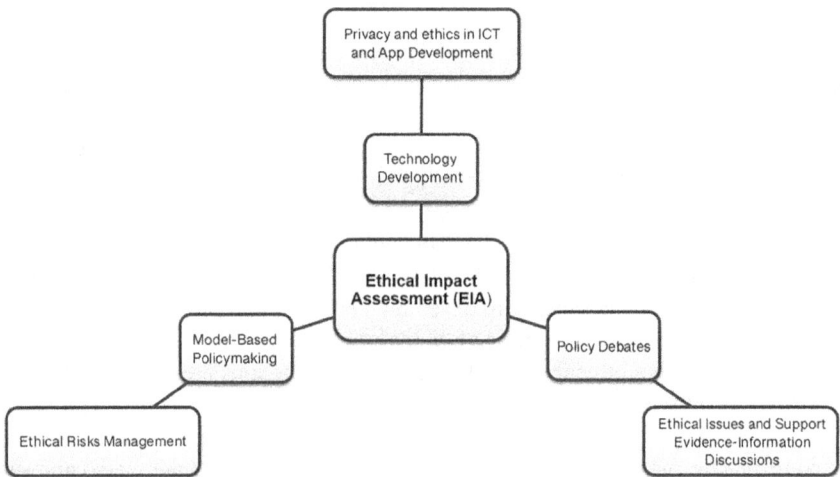

FIGURE 10.2   EIA applications

10.2.2   Stakeholder Engagement Strategies for Ethical Policy Development

Stakeholder involvement is crucial for ethical governance since it helps to guarantee that the goals and methods of innovations are socially acceptable and correspond to society's requirements and expectations [15]. It helps ensure that innovation aims and approaches are socially acceptable. The RRI internal management and procedural system is called the ETHNA System. It introduces a new official organizational framework enabling participation in the RRI process to be open and transparent. This is achieved through considering and accepting the objectives of innovations together with their methods and social acceptability.

Essential elements of successful stakeholder engagement methods are identifying significant stakeholders, learning about their requirements, expectations, and beliefs, and determining how best to include them in the decision-making procedure. The ETHNA System offers direction on how to get responses to issues such as who the pertinent stakeholders are, why they ought to be included, what they can contribute, and how to be involved most successfully [15]. One of these is the identification, by a variety of techniques, of the values, needs, and expectations of stakeholders; another is the assurance that participation comes from many directions; both of these eventually result in well-informed decisions about the choice, execution, and application of research [16].

Stakeholder involvement is necessary for creating long-term value as much as ethical policy development. Stakeholder engagement is crucial when stakeholders are involved to provide long-term value [17]. This is so because it contributes to guaranteeing the moral, strategic, and practical goals, actions, and effects of stakeholder relations. In their stakeholder connections, businesses and other organizations have used a range of processes and strategies that academics and practitioners have referred to as "stakeholder engagement." These tactics and procedures comprise working with, including, or partnering with stakeholders.

Conversely, stakeholder involvement has drawbacks that typically need to be addressed in the literature. Power imbalances, conflicts of interest, and moral dilemmas can all arise from the involvement of stakeholders, according to Andriof et al. [17], and they could compromise the effectiveness and legitimacy of the decision-making process. Therefore, it is imperative to show that integrating these two elements is essential for the progress of research on stakeholder engagement and to supplement the positive perspective common in the literature with a discussion of its dark side, which is usually ignored.

10.3 ETHICAL IMPLICATIONS OF SURVEILLANCE AND GOVERNMENTAL CONTROL

Privacy issues and ethical dilemmas are becoming more and more important in the fields of cybersecurity and new technology. The moral issues raised by cybersecurity methods are fundamental elements of digital domain ethical frameworks. These cover the need to protect user data, ethical hacking, and privacy protection. These frameworks stress principles such as explicability, beneficence, autonomy, non-maleficence, and justice to direct ethical decision-making in cybersecurity activities [18].

Surveillance and political control raise serious ethical issues regarding privacy, human autonomy, and social norms, and they also have many other unintended consequences. Artificial intelligence (AI) systems and security cameras are two examples of surveillance technology that can violate someone's right to privacy. Over-monitoring can create a surveillance state in which people believe their freedoms are restricted and they are constantly being watched. As a result, democratic countries need to find a compromise between privacy rights and security considerations. To ensure that data collection and analysis are carried out impartially and fairly, governments and tech companies using surveillance technologies also need to include accountability and transparency in their ethical frameworks [19].

Government oversight and control are crucial to lessen the risks of artificial intelligence espionage. Technology businesses and regulatory agencies working together can create and execute moral laws that uphold civil rights and improve the community's well-being. The necessity of establishing precise regulations to protect people's right to privacy while maintaining effective security measures is highlighted by the possibility that surveillance systems will be biased, discriminatory, or used maliciously. Reaching a peaceful balance between security needs and privacy concerns is essential to combating the abuse of surveillance technologies.

Predictive policing algorithms and face recognition systems are two examples of ubiquitous surveillance technology in our culture today. This emphasizes how important it is to thoroughly analyze the ethical considerations related to these kinds of technologies. Privacy invasion, bias, discrimination, and data security breaches are just a few of the moral conundrums that arise from surveillance activities.

Artificial intelligence monitoring might offer two advantages: improved security and personalized services. It also raises questions about social trust, privacy rights, and civil freedoms. To overcome the ethical difficulties surrounding artificial intelligence monitoring, parties, including governments, technology corporations, and civil society organizations, must cooperate and maintain regular communication. Prioritizing moral values can help create an atmosphere that encourages accountability and transparency and creates surveillance systems that protect people's rights, lessen discrimination, and advance society. A balance must be struck between the required security measures and the ethical issues that must be taken into account in order to protect fundamental human rights and benefit from surveillance technology at the same time.

10.3.1 ETHICAL ISSUES IN MASS SURVEILLANCE PROGRAMS

Much debate has been centered on the moral ramifications of mass monitoring systems, especially given the COVID-19 pandemic and the expanding use of artificial intelligence (AI) in surveillance [20]. Artificial intelligence monitoring has raised concerns about privacy, discrimination, and potential power abuse [20, 21]. The COVID-19 epidemic has also led to the implementation of mass monitoring measures, like health code applications, which have drawn criticism since they may cause prejudice and societal separation [20].

One of the main moral issues mass monitoring applications raise is the potential for privacy violations. Many people consider it intolerable that the collection of personal data violates privacy [19]. This is valid whether the data is accessed or utilized for deemed inappropriate activities. These problems have been made much worse by applying artificial intelligence (AI) to surveillance. Big volumes of personal data can be collected and analyzed by AI technology, often without the consent or awareness of the people being watched [21].

Concerning mass surveillance operations, the potential for power abuse raises serious ethical questions [20, 21]. Artificial intelligence monitoring has drawn criticism since it can be used for political purposes, including suppressing opposition and distorting public opinion [21]. The potential that governments will use these tools to impose illegal laws has been raised as another concern resulting from implementing mass monitoring methods during the COVID-19 pandemic [19].

10.3.2 ETHICAL LIMITS OF GOVERNMENTAL CONTROL IN CYBERSPACE

Speaking about the moral constraints of state power in cyberspace, one must consider various viewpoints and factors. It is a complex and subtle problem. The five pillars of cyber ethics emphasize the need to balance security and privacy protection: justice, nonmaleficence, explicability, beneficence, and autonomy. Among the most significant things is this. These principles stress end-to-end encrypted services, incident response, continuous monitoring and assessment of security measures, and the assurance of data confidentiality, integrity, availability, and cyber resilience. However, the quick advancement of new technologies has also brought about moral dilemmas, especially in data privacy, artificial intelligence-related risks, sustainable environments, health effects, and issues related to the weaponization of data and information [18].

Even when governments are heavily involved in creating cyberspace laws and regulations, nations usually have no agreement on the interests and priorities that should be prioritized. For example, the United States is committed to ensuring network security and freedom of access. Conversely, other countries—like China and Russia—are more worried about how access to freedom can affect their political stability. This difference in priority can make it harder to bring unlawful actors to justice because of the worldwide character of cyberspace. The U.S. Government must provide more funds to address this problem since the development of related domestic and international law has not kept up with the advancement of technology [22].

In cyberspace, the private sector is also quite important because individual companies have resources, skills, and knowledge that are on a par with or even better than their governments. The capacity of the state to impose control is naturally limited because of the uneven regulatory environment and the fact that the private sector owns and maintains cyberspace. Given this, it begs whether countries can establish their sovereignty, monopolize force, and order cyberspace to the same degree as they control their territory [23].

10.4 ENSURING ETHICAL USE OF IT IN PUBLIC SECTOR ORGANIZATIONS

The public sector's use of information technology makes ethical procurement norms imperative for ensuring fairness, accountability, and openness throughout the procurement process. These are essential steps to maintain public confidence and ensure public resources are used effectively and efficiently. As Anderson et al. [24] noted, there are significant differences in business-to-business procurement ethics and strategy between the for-profit and non-profit sectors. Corporate procurement transactions can reveal these differences. The research results show buyers' leaders in the not-for-profit sector are more likely to be prepared to overlook the opportunistic behavior of their subordinate buys. However, buyers in the for-profit sector are more likely to engage in such behavior. This suggests that non-profit organizations' procurement procedures may use some improvement.

The ethical conduct of the parties engaged mostly determines if public procurement laws are followed. According to Zadawa et al. [25], ethical behavior is necessary to lower non-compliance and attain high compliance. A prerequisite for both of these results is ethical behavior. The study's conclusions state that ethical standards must be upheld throughout procurement to prevent possibly devastating results. In agreement with this are the conclusions of Zitha and Mathebula [26], who contend that ethical behavior is a key element of procurement and may be used to restore the integrity of the government. The study results suggest that moral behavior can function as an intermediary between the variables that influence public procurement regulatory compliance and the impact of such compliance.

Public procurement officers, practitioners, and professionals should develop ethical behavior; this is a strategic problem that interests politicians and professional groups. As mentioned by Zubcic and Sims [27], practitioners and policymakers should give the development of moral principles among public procurement officers significant weight. This can be done through professional growth and the formation of an ethics code to guide their conduct, enforcement, and training while working. The study also emphasizes the requirement of procurement officers to enhance their ethical traits to achieve regulatory compliance simultaneously.

10.4.1 PROMOTING ETHICAL LEADERSHIP IN PUBLIC SECTOR IT DEPARTMENTS

Through their individual behaviors and interpersonal contacts, people who are seen as ethical leaders demonstrate normatively appropriate behavior. Through two-way communication, encouragement, and decision-making, they motivate followers to behave similarly [28, 29]. As it has been shown, ethical leadership consists of giving guidance, communicating moral standards, and fostering a sense of accountability for one's actions [28].

The relationship between moral leadership and the results that employees experience has received a great deal of attention during the last several years. As trustworthy sources of inspiration for their followers, ethical leaders encourage others to act morally and with greater initiative, which eventually helps to accomplish organizational goals [28]. Moreover, by using role modeling to motivate others, ethical leadership keeps replicating itself similarly, reinforcing itself [29].

There are several reasons why it is imperative to encourage ethical leadership in public sector information technology divisions (Figure 10.3). First, ethical leadership can help create the conditions for employee voice, highlight the inventiveness of staff members, and promote knowledge-sharing among employees. An important aspect of digital enterprises is the company's identity, which is also benefited by ethical leadership [30]. Ethical leadership is a critical element that influences firms' practices and greatly affects employee satisfaction [31].

Organizations can choose strong ethical leaders to encourage moral work conduct among employees [1]. This can be carried out to create moral leadership in information technology departments of the public sector. Moreover, companies can prioritize research and development for company expansion to maximize efficiency and gain a competitive edge over their rival companies. Emphasizing the synchronization of the relationship between the company and its employees will help achieve this. The bond between managers and their subordinates is supposed to get stronger if they keep acting with honesty, respect, and care when making decisions that affect the welfare of their staff [1]. The moral conduct of leaders of this kind inspires followers, who strengthen organizational commitment and positively impact the moral conduct of employees at work [28].

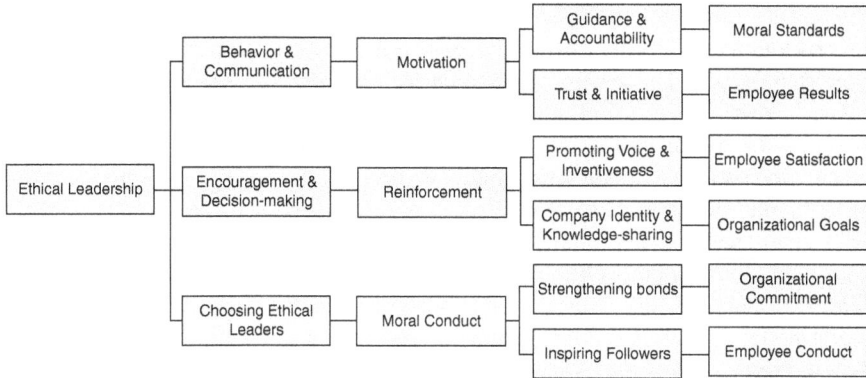

FIGURE 10.3   Keys to public sector IT department ethical leadership

10.5   ADDRESSING ETHICAL CHALLENGES IN IT REGULATION AND COMPLIANCE

Since legality, ethics, and regulations are intertwined with regulatory compliance frameworks, ethical concerns must be addressed to uphold fundamental rights and maintain public trust. The most important factors to consider are the preservation of human dignity, the honoring of personal autonomy, the assurance of justice, and prioritizing accountability and safety. To thrive in the complicated realm of artificial intelligence in healthcare, legislators, governments, system developers, and healthcare practitioners must act morally. In order to preserve people's autonomy and self-determination, this initiative attempts to create sustainable data practices, protect privacy, and give people authority over their sensitive information [32, 33]. Integrating ethical frameworks with regulatory compliance is one method to do this.

Information technology industry regulatory capture raises severe ethical questions because it can lead to legislation that prioritizes the industry's interests over the general public's welfare. This scenario, in which the businesses they oversee control or influence regulatory agencies, may lead to biased decision-making that advances corporate goals at the price of fair competition and consumer protection. The information technology industry must identify and measure regulatory capture to guarantee openness and accountability in governance. Academics stress the significance of following strict protocols to detect and prevent capture and the need to strike a balance between the interest of the public in regulatory processes and the power of industry players [34–36].

A company's voluntary efforts to fulfill its legal and ethical obligations toward society are referred to as "Corporate Social Responsibility," or CSR. In terms of information technology, this entails ensuring that corporate operations comply with the demands of regulatory bodies and society's expectations regarding issues like protecting personal data, environmental preservation, and applying fair labor standards. Corporate social responsibility (CSR) concepts integrated into regulatory compliance activities allow businesses to show a dedication to moral behavior beyond following the law. This advances society generally and fosters trust among interested parties. This category might contain projects like open reporting on the environmental effects of information technology operations, moral procurement of materials for information technology products, and inclusion and diversity promotion in the information technology workforce.

Economic penalties have been applied as a non-military enforcement force to deter and penalize non-compliance with various international accords and conventions [37]. Diverse approaches for controlling transparency have been put out in the context of public health to address and control the impact of companies on public health policy, research, and practice. Among these are education, monitoring, and identification. These systems seek to guarantee compliance, encourage transparency regarding relationships and conflicts of interest, detect, track, and inform third parties, and forbid any interaction with industry that could endanger public health [38]. Economic penalties are detrimental to health and health systems in low- and moderate-income nations. This result emphasizes the need to consider the potential for such actions to affect health [39]. Fair labor standards and worker rights are greatly advanced by ethical enforcement techniques and penalties for non-compliance [40]. The enforcement and compliance with labor legislation are two other domains in which these procedures are crucial.

10.6 ETHICAL LEADERSHIP IN SHAPING IT POLICIES AND STANDARDS

Through their individual behaviors and interpersonal contacts, people who are seen as ethical leaders demonstrate normatively appropriate behavior. Through two-way communication, encouragement, and decision-making, they motivate followers to behave similarly. When one behaves this way, others are motivated to act morally and take initiative, which eventually helps the company reach its goals. Two basic and related components of ethical leadership are creating moral principles and offering incentives, acknowledgment, and penalties to ensure these principles are followed. Leaders show good traits like honesty, integrity, trustworthiness, care, and respect for their coworkers by their deeds in the workplace [28].

Ethical leadership is a crucial element that influences employee satisfaction (ES) and is what motivates company practices. The link between ethical leadership and employee work satisfaction is strong and favorable; perceived organizational transparency (OT) and media richness (MR) mediate this connection. The use of media-based tools by ethical leaders helps to improve internal communication within the company. The more ethical leaders can influence team cohesiveness through moral instructions, and the more successful the media channel [3]. As a factor in business communication, EL helps create a moral and ethical environment. Because they are given an atmosphere in which to practice ethical behavior, the employees are empowered by the moral character of ethical leaders [30].

Ethical leadership shapes a company's culture, eventually leading to increased corporate transparency and less ambiguity. Ethical leaders can delegate authority, make clear responsibilities, give moral counsel, put people first, be aware of environmental and sustainability issues, and lead with integrity. Having these behavioral traits helps optimize the advantages of interaction between managers and staff members. In followers' views, they can establish honesty and integrity and mold a highly favorable response from staff members, including loyalty to their work [41]. A company's information technology policies and standards can only be formed with ethically grounded leadership. Encouraging moral behavior, communication, and decision-making accomplishes this, eventually leading to increased employee happiness and a positive work environment.

These traits identify ethical leaders in addition to upholding moral principles, respecting others, being just and honest, and having integrity [30, 42]. Top of their agendas includes making moral decisions, modeling moral behavior, and encouraging ethical behavior in their followers using two-way communication, reinforcement, and decision-making [30]. The potential of ethical leadership to influence employee attitudes and behaviors and encourage them to seek out business identity and match their actions with the organization's values and goals makes it crucial in information technology governance [29].

Several traits that characterize ethical leadership have been identified. Respect, following corporate values, justice, honesty and integrity, teamwork, accountability, compassionate treatment, role modeling, altruism, and crediting others are among these traits [42]. These traits are essential in building trust and fostering a positive work environment that promotes moral conduct and sound judgment. Inspiring and motivating others who follow them is another attribute that sets ethical leaders apart. They can motivate their followers to put in more effort for the company's growth and feel more satisfied at work [30].

10.6.1 BUILDING ETHICAL ORGANIZATIONAL CULTURES IN IT DEPARTMENTS

A culture that not only encourages and promotes moral behavior and decision-making but is also a necessary element of a successful corporate culture is called an ethical culture. Information technology departments must thus give the development of an ethical culture first attention.

Building an ethical culture can be accomplished by information technology departments following a multiyear, ongoing approach. This approach consists of determining the holes in the ethics culture, setting up an ethics task force, defining and ranking issues, creating a reform plan, carrying it out, and assessing the results. This process will help the information technology department to match its objectives and principles with its activities. It will also provide a basis for success and a system for daily ethical decision-making. Even in complex hierarchies, ethical leadership can significantly influence the actions of lower leaders within an organization. Information technology departments should thus give top attention to ethical leadership development. These executives need to be able to set an example of moral conduct and sound judgment.

10.6.2 ADVOCATING FOR ETHICAL STANDARDS AND BEST PRACTICES IN IT INDUSTRIES

The information technology industry must uphold ethical principles and best practices to fulfill its responsibility, considering its significant impact on society. Advocates of these ideas are necessary to ensure that the sector will be dependable, creative, and beneficial to all parties involved. In the realm of information technology, ethical standards include, for instance, promoting transparency, eliminating bias, and protecting privacy. A few best practices could be following industry standards, using safe coding methods, and always picking up new and improving abilities. Advocates for ethical standards and best practices in information technology can be found in several groups, including government statutes, professional societies, and industry associations. Governments contribute to advancing moral principles and best practices in information technology by implementing laws and regulations.

Working together and having conversations with many stakeholders—industry leaders, legislators, and the general public—is essential to effectively promoting ethical standards and best practices in information technology. Such activities include participating in policy-making processes, holding public debates, and raising awareness of ethical issues in information technology. IT experts may support ethical guidelines and best practices, helping maintain the industry as a constructive societal force. This would help guarantee that the sector promotes social welfare, economic expansion, and innovation.

REFERENCES

1. Mäntymäki, M., Minkkinen, M., Birkstedt, T. and Viljanen, M., Defining organizational AI governance. *AI and Ethics*, 2022. 2(4): pp. 603–609.
2. Munn, L., The uselessness of AI ethics. *AI and Ethics*, 2023. 3(3): pp. 869–877.
3. Gianni, R., S. Lehtinen, and M. Nieminen, Governance of responsible AI: From ethical guidelines to cooperative policies. *Frontiers in Computer Science*, 2022. 4: p. 873437.
4. Winfield, A.F. and M. Jirotka, Ethical governance is essential to building trust in robotics and artificial intelligence systems. *Philosophical Transactions of the Royal Society A: Mathematical, Physical and Engineering Sciences*, 2018. 376(2133): p. 20180085.
5. Drolet, M.J., Rose-Derouin, E., Leblanc, J.C., Ruest, M. and Williams-Jones, B., Ethical Issues in research: Perceptions of researchers, research ethics board members and research ethics experts. *Journal of Academic Ethics*, 2023. 21(2): pp. 269–292.
6. Colnerud, G., Ethical dilemmas in research in relation to ethical review: An empirical study. *Research Ethics*, 2014. 10(4): pp. 238–253.
7. Taeihagh, A., Governance of artificial intelligence. *Policy and Society*, 2021. 40(2): pp. 137–157.
8. Bouhouita-Guermech, S., P. Gogognon, and J.-C. Bélisle-Pipon, Specific challenges posed by artificial intelligence in research ethics. *Frontiers in Artificial Intelligence*, 2023. 6: p. 1149082.
9. McNamara, A., J. Smith, and E. Murphy-Hill. Does ACM's code of ethics change ethical decision-making in software development? in *Proceedings of the 2018 26th ACM Joint Meeting on European Software Engineering Conference and Symposium on the Foundations of Software Engineering*, ACM Digital Library. 2018.

10. Banks, G.C., Knapp, D.J., Lin, L., Sanders, C.S. and Grand, J.A., Ethical decision making in the 21st century: A useful framework for industrial-organizational psychologists. *Industrial and Organizational Psychology*, 2022. 15(2): pp. 220–235.

11. Madani, M., Ghasemzadeh, N., Dizani, A., Gharamaleki, A.F. and Larijani, B., Policy considerations to achieve practical ethics: Closing the gap between ethical theory and practice. *Journal of Medical Ethics and History of Medicine*, 2020. 13: pp. 1–12.

12. Mitchell, P., Reinap, M., Moat, K. and Kuchenmüller, T., An ethical analysis of policy dialogues. *Health Research Policy and Systems*, 2023. 21(1): p. 13.

13. Wright, D. and E. Mordini, Privacy and ethical impact assessment, in *Privacy impact assessment*. 2012, Springer. pp. 397–418.

14. Boden, L.A. and I.J. McKendrick, Model-based policymaking: A framework to promote ethical "good practice" in mathematical modeling for public health policymaking. *Frontiers in Public Health*, 2017. 5: p. 254465.

15. Häberlein, L. and P. Hövel, Importance and necessity of stakeholder engagement, in *Ethics and Responsible Research and Innovation in Practice: The ETHNA System Project*. 2023, Springer. pp. 38–53.

16. Petkovic, J., Riddle, A., Akl, E.A., Khabsa, J., Lytvyn, L., Atwere, P., Campbell, P., Chalkidou, K., Chang, S.M., Crowe, S. and Dans, L., Protocol for the development of guidance for stakeholder engagement in health and healthcare guideline development and implementation. *Systematic Reviews*, 2020. 9: pp. 1–11.

17. Andriof, J., Rahman, S.S., Waddock, S. and Husted, B., Introduction: JCC theme issue: Stakeholder responsibility. *The Journal of Corporate Citizenship*, 2002: pp. 16–19.

18. Dhirani, L.L., Mukhtiar, N., Chowdhry, B.S. and Newe, T., Ethical dilemmas and privacy issues in emerging technologies: A review. *Sensors*, 2023. 23(3): p. 1151.

19. Königs, P., Government surveillance, privacy, and legitimacy. *Philosophy & Technology*, 2022. 35(1): p. 8.

20. Boersma, K., M. Büscher, and C. Fonio, Crisis management, surveillance, and digital ethics in the COVID-19 era. *Journal of Contingencies and Crisis Management*, 2022. 30(1): pp. 2–9.

21. Saheb, T., Ethically contentious aspects of artificial intelligence surveillance: A social science perspective. *AI and Ethics*, 2023. 3(2): pp. 369–379.

22. Trujillo, C., The limits of cyberspace deterrence. *Joint Force Quarterly*, 2014. 75(4): pp. 43–52.

23. Hoffman, W. and S. Nyikos, *Governing private sector self-help in cyberspace: Analogies from the physical world*. 2022: Carnegie Endowment for International Peace.

24. Hawkins, T.G., M.J. Gravier, and E.H. Powley, Public versus private sector procurement ethics and strategy: What each sector can learn from the other. *Journal of Business Ethics*, 2011. 103: pp. 567–586.

25. Zadawa, A.N., A.A. Hussin, and A. Osmadi, Determinants of compliance with public procurement guidelines in the Nigerian construction industry. *Jurnal Teknologi*, 2015. 75(9): pp. 107–110.

26. Zitha, H. and N. Mathebula, Ethical conduct of procurement officials and implications on service delivery: A case study of Limpopo provincial treasury. *Public and Municipal Finance*, 2015. 4(3): pp. 16–24.

27. Zubcic, J. and R. Sims, Examining the link between enforcement activity and corporate compliance by Australian companies and the implications for regulators. *International Journal of Law and Management*, 2011. 53(4): pp. 299–308.

28. Guo, F., Xue, Z., He, J. and Yasmin, F., Ethical leadership and workplace behavior in the education sector: The implications of employees' ethical work behavior. *Frontiers in Psychology*, 2023. 13: p. 1040000.

29. Hosseini, E. and J.J. Ferreira, The impact of ethical leadership on organizational identity in digital startups: Does employee voice matter? *Asian Journal of Business Ethics*, 2023. 12(2): pp. 369–393.

30. Guo, K., The relationship between ethical leadership and employee job satisfaction: The mediating role of media richness and perceived organizational transparency. *Frontiers in Psychology*, 2022. 13: p. 885515.

31. Zahari, A.I., Said, J., Muhamad, N. and Ramly, S.M., Ethical culture and leadership for sustainability and governance in public sector organisations within the ESG framework. *Journal of Open Innovation: Technology, Market, and Complexity*, 2024. 10(1): p. 100219.

32. Mennella, C., Maniscalco, U., De Pietro, G. and Esposito, M., Ethical and regulatory challenges of AI technologies in healthcare: A narrative review. *Heliyon*, 2024. 10(4): pp. 1–20.

33. Kisselburgh, L. and J. Beever, The ethics of privacy in research and design: Principles, practices, and potential, in *Modern socio-technical perspectives on privacy*. 2022, Springer International Publishing Cham. pp. 395–426.

34. Li, W.Y., Regulatory capture's third face of power. *Socio-Economic Review*, 2023. 21(2): pp. 1217–1245.

35. Tu, Y., Peng, B., Elahi, E. and Wu, W., Initiator or intermediary? A case study on network relation of environmental regulatory capture in China. *International Journal of Environmental Research and Public Health*, 2020. 17(24): p. 9152.

36. Dal Bó, E., Regulatory capture: A review. *Oxford Review of Economic Policy*, 2006. 22(2): pp. 203–225.

37. Erickson, J.L., Punishing the violators? Arms embargoes and economic sanctions as tools of norm enforcement. *Review of International Studies*, 2020. 46(1): pp. 96–120.

38. Mialon, M., Vandevijvere, S., Carriedo-Lutzenkirchen, A., Bero, L., Gomes, F., Petticrew, M., McKee, M., Stuckler, D. and Sacks, G., Mechanisms for addressing and managing the influence of corporations on public health policy, research and practice: A scoping review. *BMJ Open*, 2020. 10(7): p. e034082.

39. Pintor, M.P., M. Suhrcke, and C. Hamelmann, The impact of economic sanctions on health and health systems in low-income and middle-income countries: A systematic review and narrative synthesis. *BMJ Global Health*, 2023. 8(2): p. e010968.

40. Syed, R.F., Compliance with and enforcement mechanism of labor law: Cost-benefits analysis from employers' perspective in Bangladesh. *Asian Journal of Business Ethics*, 2023. 12(2): pp. 395–418.

41. Metwally, D., Ruiz-Palomino, P., Metwally, M. and Gartzia, L., How ethical leadership shapes employees' readiness to change: The mediating role of an organizational culture of effectiveness. *Frontiers in Psychology*, 2019. 10: p. 434635.

42. Shahab, H., Zahur, H., Akhtar, N. and Rashid, S., Characteristics of ethical leadership: Themes identification through convergent parallel mixed method design from the Pakistan context. *Frontiers in Psychology*, 2021. 12: p. 787796.

11 Risk, Responsibility, and Ethical Accountability in IT

11.1 RISK IN IT AND ITS ETHICAL IMPLICATIONS

Information technology (IT) is associated with numerous types of risks, including those about privacy and security, those pertaining to technology abuse, and those pertaining to unanticipated consequences of technology. As more private and delicate data is kept online, there is a higher risk of unauthorized access, theft, or misuse of this data. The more digital storage one has, the higher this danger. This may have serious consequences for people, such as identity theft, money loss, or reputational harm. Applying strict security measures and data protection procedures is the ethical duty of IT companies and experts to safely protect sensitive information and maintain user confidence [1].

Algorithms and machine learning models may inadvertently mirror and perpetuate societal prejudices found in the data on which they are trained. Biased results may follow in industries like lending, criminal justice, or recruiting [2]. It is incumbent upon IT workers to be vigilant in detecting and reducing these prejudices. This will guarantee ethical and just development and application of technology.

The speed at which technology develops raises ethical questions about how information technology may affect jobs and the workforce. Artificial intelligence (AI) and automation systems are getting increasingly sophisticated, which raises the risk of job losses and even more noticeable economic inequality [3]. IT professionals and legislators must consider the moral ramifications of these developments and start developing solutions that balance technology development with the stability of the economy and the welfare of individuals and communities. Today, businesses operate where technology is the driving force behind many risks related to their digital assets and IT infrastructure. Knowing more about the nature of these risks is crucial to creating effective risk management strategies that uphold moral standards and avoid possible harm. As Figure 11.1 shows, there are several ways to categorize the different kinds of IT hazards.

11.1.1 Impact of IT Risks on Individuals and Society

Particularly among younger users, excessive internet use has been connected to psychological problems like anxiety, depression, and low self-esteem. People may also be exposed to offensive material, cyberbullying, and online harassment due to the anonymity and lax regulations of the internet, which can have detrimental psychological effects [4]. People and society are at serious risk due to spreading false and misleading information online. People's ideas and behaviors can be influenced by misinformation, which can result in bad choices and sometimes dangerous behavior. For instance, the COVID-19 pandemic's dissemination of false health information weakened public confidence in scientific institutions and impeded efforts to contain the virus [5]. IT hazards can also significantly impact society, escalating already-existing disparities and fostering the emergence of authoritarianism. The unequal distribution of technology access and digital knowledge, known as the "digital divide," can exacerbate the marginalization of underprivileged populations and reduce their chances of achieving social and economic success. Worries about privacy violations and the loss of civil liberties are raised by the concentration of power in a small number of powerful digital businesses and the possible misuse of personal data.

AI is increasingly being used in the workplace, which brings new hazards that may make it harder for people to have meaningful work experiences. AI-driven monitoring and surveillance systems have the potential to limit worker autonomy, which might result in dishonest behavior and a feeling of control. Concerns regarding job displacement and the requirement for retraining and reskilling are also raised by the possibility that AI would eventually replace human labor in some tasks [6].

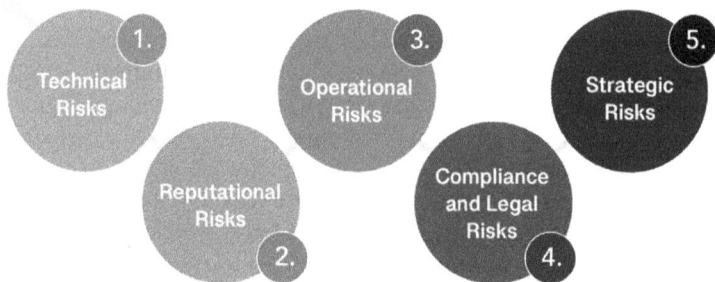

FIGURE 11.1   Classification of IT risks

11.1.2   ETHICAL IMPLICATIONS OF DATA PRIVACY AND SECURITY RISKS

Serious consequences of failing to preserve personal data include discrimination, financial loss, reputational harm, and violation of individual autonomy and privacy rights [7, 8]. To be ethical, researchers must ensure adequate permission procedures, protect disadvantaged groups, advance equity, and reduce risks to research participants [9]. Conversely, the rules now in place often need to be better defined, with unclear guidelines for data ownership, sharing, and use [7, 9].

Using personal data poses issues with social justice, fairness, and collective privacy in addition to individual privacy. Communities that are already marginalized can be stigmatized and exploited in part by bias in algorithms and the re-identification of anonymized data. The power disparity between citizens, governments, and technology companies is another matter that needs to be resolved. Important public involvement and a more thorough ethical framework are needed to guarantee that the benefits of data-driven technologies are provided equitably [9].

Prerequisites for legislators are the creation of impartial monitoring bodies and the adoption of thorough data protection laws. Businesses should go beyond simply following the law in carrying out their corporate social responsibility and instead implement moral data practices. Data ethics ultimately involves building a fair, trustworthy, and just data ecosystem that puts the preservation of human rights and the progress of the common good first rather than only safeguarding people's privacy.

11.2 ETHICAL RESPONSIBILITIES IN RISK ASSESSMENT AND MANAGEMENT

To play this function, accountability and openness must be maintained. IT professionals are obligated to ensure that all pertinent parties are informed of the possible risks associated with technological systems and processes. To do this, it is imperative to thoroughly identify and assess risks and effectively inform the relevant stakeholders about them. Clear risk assessment procedures enable stakeholders to decide with knowledge whether to adopt and use IT solutions. These rulings follow the ideas of fairness and autonomy.

Ethics risk assessment and management need inclusion and diversity, which are fundamental to decision-making processes. IT teams should prioritize including people with various experiences and backgrounds since many points of view may lead to a more comprehensive risk assessment. When organizations include a variety of viewpoints in their strategic planning, they can identify blind spots, challenge presumptions, and raise the standard of risk management plans. A culture where ethical issues are included in every aspect of IT operations is fostered by inclusion, promoting a sense of ownership and shared responsibility.

Achieving a balance between the many stakeholder interests is another aspect of ethical risk management. Even if businesses might give financial gains or operational effectiveness more weight, they still have to consider how their activities will affect a range of stakeholders, including staff, clients, and the community in which they operate. IT workers have to carefully weigh the conflicting interests at work and strive to reduce risks in a manner that respects the rights and welfare of all parties concerned to carry out their ethical obligations. Businesses may handle complex risk environments and yet adhere to the values of social responsibility and integrity if they use moral decision-making processes and get input from a broad range of stakeholders.

11.2.1 TRANSPARENCY AND ACCOUNTABILITY IN RISK ASSESSMENT

It is important to be transparent when discussing risks, their evaluation, and the decision-making procedure. The public and other stakeholders can then comprehend and evaluate the risk assessment. We need to be transparent to establish trust and make educated decisions [10–12].

Regarding risk assessment, accountability means that those in charge of the process are made to answer for their choices and actions. Methods for supervision, redress, and specific duties and decision-making procedures must be laid out. Thorough and public-interested risk assessment can be achieved through accountability [10, 12, 13].

However, being completely open and accountable can be challenging, which is not always easy. The usefulness of disclosures is limited since they are frequently unavailable, inconsistent, or incomplete [14]. Transparency and accountability can be hindered by factors such as organizational cultures, power dynamics, and conflicts of interest. A combination of measures, such as standardized reporting, independent supervision, and a new attitude toward transparency and education, is needed to solve these problems [10, 11, 13].

11.2.2 INCLUSIVITY AND DIVERSITY IN RISK MANAGEMENT TEAMS

Teams with members from various demographics and life experiences can better spot and respond to threats [15, 16]. Emerging trends, unique client needs, and possible crises can be better identified by teams that reflect the diversity of the industry and customer base. Teams with various perspectives can better anticipate and plan for unforeseen challenges [15]. Inclusion and diversity in risk management teams require representation, diversified knowledge, cultural competency, collaboration, and constant learning.

In times of crisis, the resolution of problems might be impeded by potential obstacles such as interpersonal disputes and the inability to reconcile different viewpoints. Diversity management strategies can reduce these risks by prioritizing inclusive leadership, creating supportive team climates, and implementing diversity-focused HR policies [16]. Managers are responsible for fostering an inclusive work environment and should face consequences for failing to meet diversity and inclusion goals.

Following a growth mindset and addressing inherent biases in sponsor-protégé interactions, these programs should be open to all employees. Staff members from underrepresented groups can better cope with bias and discrimination if they can access development opportunities tailored to their needs.

11.2.3 BALANCING STAKEHOLDER INTERESTS IN RISK MITIGATION STRATEGIES

Ranking stakeholders based on their power, influence, and the influence the organization has on them is one of the most crucial elements. Organizations can create tailored engagement strategies for any group, whether they are low-power/low-interest stakeholders who might need to be informed or high-power/high-interest stakeholders who need close cooperation [17]. By "stakeholder mapping," organizations can create these plans.

Crucially, the involvement of stakeholders needs to be a continual process of mutual understanding and constructive decision-making rather than a one-time event. Scientific knowledge, reasoned decision-making, and public values are partially reconciled by this "analytic-deliberative discourse." Companies can develop risk mitigation plans that are more thorough but more credible and successful over time by incorporating a variety of viewpoints [18].

Companies must proactively recognize emerging issues from stakeholders and modify their approach appropriately. Beyond lowering risks, "opportunities analysis" describes finding new ways stakeholders may provide value. Companies that embrace inclusive governance and find a balance between the competing interests of their stakeholders can strengthen their "license to operate" and guarantee the sustainability of their risk management initiatives [19].

11.3 BALANCING ORGANIZATIONAL GOALS WITH ETHICAL RISK MITIGATION

Modern businesses need help reconciling the company's goals with minimizing ethical risks. On the one side, these companies must follow their strategic goals to stay competitive and successful. Still, this has to be done morally justly, lowering the risks to customers, workers, and society at large [20, 21].

Including moral principles in the safety management systems and corporate governance of the relevant company is one of the most crucial elements. Within this is creating an ethical management policy that shows a strong dedication to ethics and safety throughout the company. Moral leadership is important because it leads by example and prioritizes safety. Companies should create an ethical atmosphere where every employee has the information required to conduct business morally [22].

11.3.1 ALIGNING BUSINESS OBJECTIVES WITH ETHICAL RISK TOLERANCE

This requires the organization to include its mission, values, and environmental, social, and governance (ESG) concerns in its business strategy and operations [23, 24]. To ensure that they comply with legislation and industry-wide voluntary ESG norms, companies must have strict rules, standards, due diligence procedures, and internal audit systems [24].

Setting clear and deadlined goals, targets, and KPIs is essential to guarantee accountability and transparency when managing shared risks and generating shared benefits. These objectives and targets should be arranged to be assessed objectively or subjectively if companies want to report publicly on their performance. Business enterprises should also conduct thorough data collecting, assessment, and analysis. This can lead to better corporate decisions, managers being held responsible and rewarded, and significant stakeholders' confidence can be established and preserved [24].

Research indicates that managers' "tone from the top" can help create a "risk-aware" culture by setting expectations for acceptable risk and empowering staff members to express their safety concerns. This may motivate employees to raise safety concerns, particularly in unclear and complicated situations like the COVID-19 epidemic. Improved risk perception is associated with ethical leadership and can encourage staff members to raise safety issues. This fact emphasizes ethical leadership's need to ensure that risks are understood and taken in the best interests of organizational members in a wide range of work contexts and situations [25].

11.3.2 Integrating Ethical Risk Management into Strategic Planning

Ethically informed risk management is the management of general hazards and ethical risks, such as those related to advance directives and the disclosure of unintended harm. To carry out their moral obligations, risk managers in the healthcare sector must practice an evidence-based, patient-centered risk management plan. This entails reducing unproductive practices, putting proactive disclosure and settlement regulations into place, and implementing proven risk control strategies. An official code of ethics for risk managers could also counter pressure from administrators or physicians to lower ethical standards and offer clarity in challenging ethical situations. One way to do this would be to create an official code of ethics.

Integrating ethical issues into corporate risk management and strategy planning outside the healthcare industry can help companies advance their moral convictions, strengthen their relationships with stakeholders, and build resilience. This demands a risk management approach that is more strategic and complex and considers the impact risk has on the viability of the business [26]. Including ethicists in creating new technologies, such as AI for medical applications, can also help foresee and address ethical issues right from the start.

11.3.3 Managing Conflicts of Interest in Risk Mitigation Strategies

Conflicts of interest can provide several major challenges in creating risk mitigation strategies. In healthcare organizations, risk managers have to balance the need to safeguard the institution with honoring the needs of the patients and clients they serve. Institutional conflicts of interest may arise when the risk manager's goal of resolving possibly costly claims on conditions advantageous to the company conflicts with the patient's best interests.

Disclosing openly and honestly is one of the most crucial ways to settle these conflicts. One study's interviewee underlined the need to disclose any conflicts of interest that could develop while planning research, requesting grants, and submitting findings for publication. All those concerned can assess whether or not possible conflicts of interest are acceptable when made visible. However, disclosure alone cannot be enough since it can provide the false impression that a contentious dispute is acceptable.

Managing conflicts of interest inside institutions depends heavily on organizational norms and legal frameworks. Professional experts suggest four ways to tackle the issue: assessing similar interests, divergence and regulatory planning, applicability, and an exit strategy. Institutions should also set aside money so that researchers may continue their work if they have to end a partnership that is giving them trouble. Decisions with ambiguous cutoff points and prone to bias should be evaluated outside of the company by an unbiased third party [27].

Apart from financial conflicts, experts note that non-commercial conflicts of interest, such as those involving government funders, can be just as challenging to handle, if not more so. Those questioned gave instances of times when government organizations tried to change trial designs or stop the release of unfavorable findings [28]. As such, it could be required to reconsider the part that healthcare professionals play as members of a market society in order to solve these structural issues. Table 11.1 lists common IT risk mitigation conflicts of interest, their dangers, and recommended mitigation solutions.

TABLE 11.1
IT risk mitigation strategies for handling conflicts of interest

Conflict of Interest	Example	Risks	Mitigation
Vendor Relationships	IT manager selects a security solution from a vendor they have a personal relationship with.	Risk of bias toward favored vendors compromised solution quality.	Implementation of strict vendor evaluation criteria, relationship transparency, and role rotation.
Insider Trading	An IT employee trades stocks based on non-public information about upcoming cybersecurity initiatives.	Breach of trust, legal consequences, erosion of morale.	Enforcement of strict trading policies, robust monitoring, and comprehensive training.
Personal Bias	The project manager favors a strategy that aligns with personal preferences and ignores evidence.	Overlooked options, reduced objectivity.	Promotion of diversity, formation of cross-functional teams, adoption of structured decision-making processes.
Financial Incentives	CTO prioritizes cost-saving over robust risk management for short-term gain.	Compromised security, legal liabilities.	Align incentives, adopt proactive metrics, and conduct regular audits.
Client Relationships	Cybersecurity consultant favors a solution benefiting their vendor affiliate over client interests.	Biased advice, reputational damage.	Definition of boundaries, establishment of explicit agreements, maintenance of independence.

11.4 ETHICAL CHALLENGES IN DATA SECURITY AND BREACH RESPONSE

Composing sensitive information carries significant consequences for personal privacy, trust, and possible harm. Researchers have determined that respect for people, beneficence, and justice are among the fundamental ethical precepts that ought to direct data breach reaction [29, 30].

Respect for persons dictates that people should be seen as autonomous agents and be free to decide what personal data to use based on correct information. This means, should a breach occur, providing affected parties with prompt and clear notice and a description of the nature of the incident and any associated risks [29]. Companies must show proof that they have taken the necessary steps to protect data and reduce the chance of damage, so accountability and transparency are crucial.

According to the beneficence concept, organizations have to maximize the benefits to those whose data has been compromised and reduce the possible risks to those people. This entails offering afflicted clients substitute remedial services such as identity theft protection, credit monitoring, and others [3]. Companies are required to do thorough research to understand the scope and kind of the breach and to take action to stop future incidents of this kind.

A prerequisite of justice is an equitable allocation of the benefits and costs related to data collection and use. This suggests that, in the case of a breach, companies are obligated to ensure that all affected parties are treated fairly, independent of the particular characteristics or circumstances [30]. It requires businesses to provide sufficient reparations and compensation to those harmed by the confidentiality violation. Businesses must evaluate how data breaches and security incidents affect society at large. Breaches might damage the economy and national security and undermine public trust in organizations and technical systems.

11.4.1 Proactive Measures for Ethical Data Security

Adopting privacy by design and privacy by default principles is a crucial proactive step that needs to be done in the development of digital technologies. In order to do this, privacy considerations are integrated across a system's whole lifecycle, starting with its design and continuing through its implementation. This ensures that privacy is the default setting and reduces privacy concerns.

Businesses should carefully assess their dynamic information asset inventory, the aspects of their data critical to realizing their full potential, and the standards by which they will judge the worth of their data and the returns on their investments. Companies need to take proactive steps by proactively recognizing and managing the risks related to privacy to protect individual privacy and avoid data breaches.

Enforcing privacy laws proactively is necessary and crucial for encouraging moral data security practices. Regulations should be proactive in enforcing both the privacy policies' requirements and the management practices. The concepts of privacy by default and privacy by design ought to be incorporated into these procedures. Apart from merely documenting incidents of noncompliance or infractions, an organization's data protection behavior can offer valuable insights into the effectiveness of the regulatory framework.

To develop a culture of ethical data protection, businesses must undertake proactive staff training and awareness campaigns. To guarantee that security controls remain effective and to eliminate risks, staff should get training on secure operations and procedures. Giving employees the freedom to independently detect and address data security issues can help organizations strengthen their overall cybersecurity posture and safeguard sensitive data from potential attackers.

11.4.2 Ethical Considerations in Data Collection and Storage Practices

Ethical standards are crucial in data gathering and storage, especially when handling sensitive data. Three basic ethical principles are beneficence (minimizing harm and optimizing benefit), justice (ensuring an equal distribution of research responsibilities), and autonomy (getting informed consent) (Figure 11.2) [31].

Safeguarding the participants' privacy and confidentiality is one of the special ethical considerations. Researchers should avoid collecting unnecessary personal information and provide the data anonymously [32]. Using pseudonyms, removing identifiers, and putting safe data storage methods in place are all crucial strategies [33]. It is crucial that consent be granted after receiving all relevant information, and researchers should be transparent about how the data will be handled and stored [34]. Verbal consent is preferable to written consent in delicate situations [33].

Ethical review procedures are used to make sure that research projects follow the relevant ethical standards. A simple review might be carried out for non-experimental projects without participants, but if participants would receive compensation, a more thorough assessment might be carried out. The hazards involved may determine the different levels of examination. Researchers should give thorough and extensive documentation on their plans for gathering data, the risks to participants, and the steps they took to ensure confidentiality [32].

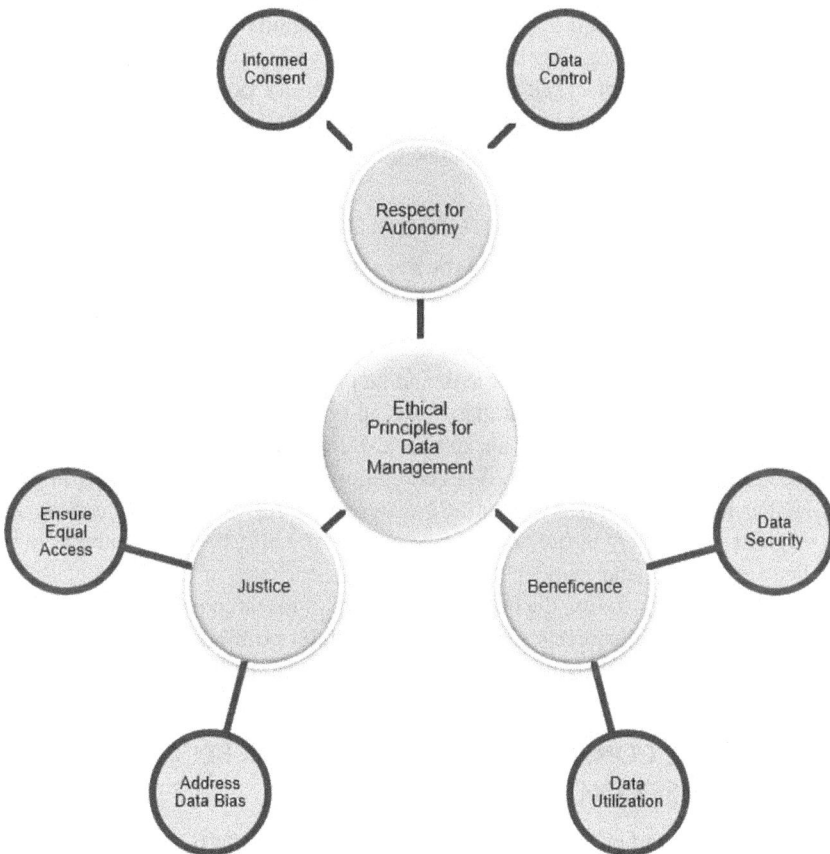

FIGURE 11.2 Principles of ethics for data gathering and storage

11.4.3 Responsiveness and Transparency in Data Breach Notification

For businesses to efficiently detect and communicate regarding vulnerabilities, well-defined processes and systems are essential. The concept of "transparency" pertains to being candid and forthright regarding a data breach, wherein pertinent stakeholders are provided with current information regarding the incident's nature, the potential severity of the repercussions, and the corrective actions implemented.

It is incumbent upon organizations to maintain transparency in their interactions with the general public, industry stakeholders, regulatory agencies, and affected individuals. Through its candid communication, the organization demonstrates its commitment to responsibility. It instills trust in its stakeholders that it handles the breach with the utmost gravity it merits and makes every effort to mitigate the impact. Failure to disclose a security incident transparently may exacerbate skepticism and damage the organization's reputation, leading to consequences beyond the immediate issue.

In addition to promptly notifying stakeholders and individuals impacted by the breach, responsiveness requires providing them with guidance and support. Support in managing the fiscal and legal consequences of the breach could be provided, in addition to resources for credit monitoring and identity protection. Organizations can demonstrate their commitment to ethical conduct and assume accountability for the consequences of a security compromise through proactive assistance extended to affected parties.

11.4.4 ETHICAL OBLIGATIONS IN DATA BREACH REMEDIATION AND RECOVERY

To ensure transparency and support their use of the data, researchers should describe these specifics in their research. Researchers should also ascertain whether any laws were breached during the data leak and be prepared to justify using the data in light of this [35]. Secondly, researchers have an obligation to mitigate any possible harm that can result from using hacked data. Before using their data, people who have been impacted must consent. It is also important to carefully consider how the data-gathering process and any resulting biases may affect the research's conclusions [29, 35]. For researchers to use compromised data ethically, they must adhere to the beneficence, justice, and respect for human rights outlined in the Belmont Report [35]. This means prioritizing the health and welfare of research participants above all else, making sure that subjects are chosen fairly, and optimizing the benefits of the study while minimizing any hazards or harm that might result from it.

11.5 ETHICAL CONSIDERATIONS IN DISASTER RECOVERY AND BUSINESS CONTINUITY

Disaster recovery and business continuity ethics include fairness, equity, and human dignity. Ethical disaster recovery is essential. This includes safeguarding the organization, employees, customers, and disaster victims. Ethical and socially responsible rehabilitation requires careful consideration of these objectives.

Disaster and business continuity ethics necessitate fairness and equity in recovery. Disasters hit certain populations more, worsening inequality. Ethical recovery planning should prioritize vulnerable populations like low-income communities, minorities, and persons with disabilities to ensure they have enough assistance and resources. This may involve fair resource distribution, specialized assistance, and systemic recovery barriers.

Crisis recovery and firm continuity depend on ethical decision-making. Leaders and decision-makers must decide complex ethical dilemmas under immense pressure and uncertainty in crises. Ethical decision-making requires transparency, honesty, and integrity. Leaders must prioritize stakeholders and act ethically when making tough decisions. Diversity and ethical thinking can also help ensure ethical decisions.

Human dignity and rights are essential for ethical disaster recovery and economic continuation. Destruction can displace communities and hurt people emotionally and physically. Safe, private, and autonomous human rights should guide ethical healing. Treat victims with decency and respect, support their rights during recovery, and avoid behaviors that could increase their vulnerability or suffering.

Recovery from disasters has long-term ethical implications. Organizations must consider how their ethics affect future generations, the environment, and societal resilience. Sustainable recovery, environmental preservation, and disaster resilience and well-being are included. Addressing ethics during disaster recovery and business continuity can show ethical stewardship and develop more resilient and equitable societies.

11.5.1 Ensuring Fairness and Equity in Business Continuity Measures

The World Trade Organization (WTO)'s pluralist approach to trade agreements offers many perspectives that can be applied to balance trade liberalization and regulatory flexibility [36]. Differentiated commitments and unilateral trade remedies are permitted under the WTO's structure to recognize member states' different social and economic governance models. Similarly, business continuity management strategies should consider the unique needs and constraints of different stakeholder groups to provide inclusive and equitable outcomes.

The concept of financial sustainability offers a useful angle that may be applied when assessing the long-term viability of business continuity plans [37]. Moreover, actions that consider the stakeholder interests, the risk-adjusted cost of capital, and short-term stock price volatility can help guarantee the sustainability and equitable application of crisis response strategies [38].

11.5.2 Long-Term Ethical Implications of Disaster Recovery Efforts

Short-term needs, typically the focus of disaster recovery efforts, include the supply of shelter, the restoration of infrastructure, and the resumption of business operations. Nonetheless, consideration must be given to the potential long-term ethical implications of healing decisions. To ensure that recovery efforts will enhance residents' quality of life, organize the safe use of land and infrastructure, and attend to everyone's needs and concerns, inclusive discussion is necessary at all governmental levels and with the general public.

Pre-existing vulnerabilities and inequities may be made worse by catastrophe recovery if it is not handled carefully. Numerous factors, including socioeconomic circumstances, historical background, and resource accessibility, affect how quickly different communities recover. Recovery policies must be evaluated to see if they can worsen already-existing power imbalances and address long-term repercussions. Particularly in international settings, ethical catastrophe management must consider the likelihood that victims may adhere to different ethical traditions and standards of justice than those who respond to the tragedy [39].

The triage procedure and the distribution of limited resources provide serious ethical challenges in disaster response. The concepts of utility, justice, fairness, and equity, and the arguments supporting the triage criteria and procedures must all be carefully considered. Topics, including the duty of care, moral suffering, professional standards, reciprocity, and societal worth, are covered by catastrophe ethics. Conversely, virtue-based ethics emphasizes doctors' obligations to treat patients, while utilitarian approaches focus on maximizing survival with limited resources.

11.6 PROMOTING A CULTURE OF ETHICAL ACCOUNTABILITY IN IT ORGANIZATIONS

One of the most crucial elements is establishing clear ethical guidelines and standards that direct the creation, advancement, and application of technology within the organization. It is the duty of businesses to take preemptive action to foster an ethical workplace. These steps include putting in place thorough training programs that instruct staff members on making moral decisions and giving them the voice to express concerns about possible abuse or unexpected consequences of the technologies they use. Leaders and managers should also set an example of ethical behavior and function as role models because their decisions and actions can significantly affect the organization's ethical culture. To ensure that their technology systems are operating as intended, businesses should set up responsible and transparent procedures. Frequent audits should be part of these protocols to find any biases or other issues that can jeopardize justice and the interests of society. Suppose IT companies embrace an ethical responsibility culture. In that case, they may help ensure that the technologies they create and use align with responsible innovation principles and benefit society.

11.6.1 LEADERSHIP'S ROLE IN CULTIVATING ETHICAL CULTURE

When leaders regularly act ethically and respect ethical values in their relationships with stakeholders, clients, and employees, a foundation for an ethical culture can be laid. This requires maintaining moral standards on one's own, ensuring accountability on the part of others, and building an integrity-focused culture inside the organization. This requires holding other individuals accountable for their actions.

Effective leadership understands the value of communication in developing a moral culture. They have an honest and open discussion with employees about the organization's ethical standards, business values, and the significance of moral decision-making in IT. Leaders help their team members make ethically sound decisions by fostering debate about moral quandaries and assisting in resolving them.

Furthermore, leaders actively seek input from staff members on moral matters, displaying a commitment to diversity and ensuring that various perspectives are considered when making moral judgments.

Executives are responsible for setting a positive example, encouraging open communication, and developing defined ethical norms and processes within the organization. Creating a code of conduct that describes anticipated behaviors and processes for reporting ethical problems or infractions is an important initial step. An organization's leaders lay the groundwork for ethical decision-making at all levels of the organization by clearly defining ethical behavior and the proper course of action to take when confronted with ethical quandaries.

Leadership is essential in developing an ethical culture beyond obeying rules and regulations. It entails creating an atmosphere where employees can raise ethical concerns without fear of penalties. Effective leaders foster a culture of psychological safety in which people feel empowered to express their thoughts, ask questions, and criticize unethical behavior or practices. Leaders can demonstrate their commitment to building a culture of trust and accountability inside the organization by requesting employee feedback and immediately pleasantly resolving issues.

11.6.2 Fostering Ethical Awareness and Education among IT Professionals

Cybersecurity professionals face many moral and ethical dilemmas when trying to protect sensitive data, thwart malicious attacks, and handle various security events. However, in this profession, there often needs to be explicit regulatory rules to ensure universal ethical standards. Ethics must unquestionably be included in cybersecurity and IT education. A larger number of individuals could benefit from AI if ethics related to the technology were taught to nursing students, for example, as this would help lessen its consequences [40].

Employees in IT should have constant access to training and resources regarding ethical issues. Workers who are more knowledgeable about cybersecurity and IT will also be more aware of these industries' ethical issues [41]. Information about the applications and advantages of artificial intelligence technology can also help to increase acceptance of AI-based technologies and the intention to employ them [40].

If IT professionals create clear ethical guidelines and standards, they may find it easier to manage challenging circumstances. For instance, a paper on the possible benefits and moral dilemmas of AI was recently published by the European Union (EU). The study emphasizes the importance of appropriate laws, programs, and research to get positive outcomes [40]. An ethical decision-making roadmap can be developed with comparable guidelines tailored to specific IT domains.

Promoting and enabling open dialogue and cooperation between IT specialists, ethicists, and stakeholders can help raise ethical awareness. Through experience sharing, case study discussion, and solution-focused teamwork, ethical concerns can be identified and addressed more effectively [42]. Encouraging the IT community to establish forums for dialogue and knowledge sharing is important.

IT professionals who demonstrate good ethical conduct or behavior can be recognized and rewarded, which helps to raise awareness of the importance of ethical behavior. Adopting recognition and incentive systems emphasizing moral decision-making and actions can encourage professionals to prioritize ethics in their jobs. This could inspire others to follow suit and create a culture of ethical awareness inside the IT sector.

11.6.3 REWARDING ETHICAL BEHAVIOR AND ACCOUNTABILITY IN IT ORGANIZATIONS

Leaders who demonstrate moral behavior and hold their staff to high ethical standards can foster a more ethical culture [43]. Leaders may stress the significance of ethical conduct by laying out standards for staff, offering ethical training, and rewarding those who do the right thing. It is equally important to be held accountable. In order to keep an eye out for and deal with unethical behavior, IT companies should set up solid processes. Clear disciplinary processes for ethical transgressions, frequent audits, and whistleblower hotlines may fall under this category [43, 44]. An organization's dedication to ethical procedures is shown when all employees, including leadership, are held accountable for their conduct. Crucial component number two is rewarding ethical behavior. Companies in the IT industry should consider how they may include ethical behavior in their performance reviews and pay scales.

REFERENCES

1. Card, A.J., What is ethically informed risk management? *AMA Journal of Ethics*, 2020. 22(11): pp. 965–975.
2. Stahl, B.C. and D. Eke, The ethics of ChatGPT–Exploring the ethical issues of an emerging technology. *International Journal of Information Management*, 2024. 74: p. 102700.
3. Howe III, E.G. and F. Elenberg, Ethical challenges posed by big data. *Innovations in Clinical Neuroscience*, 2020. 17(10–12): p. 24.
4. Diomidous, M., Chardalias, K., Magita, A., Koutonias, P., Panagiotopoulou, P. and Mantas, J. Social and psychological effects of the internet use. *Acta Informatica Medica*, 2016. 24(1): p. 66.
5. Ecker, U.K., Lewandowsky, S., Cook, J., Schmid, P., Fazio, L.K., Brashier, N., Kendeou, P., Vraga, E.K. and Amazeen, M.A., The psychological drivers of misinformation belief and its resistance to correction. *Nature Reviews Psychology*, 2022. 1(1): pp. 13–29.
6. Bankins, S. and P. Formosa, The ethical implications of artificial intelligence (AI) for meaningful work. *Journal of Business Ethics*, 2023. 185(4): pp. 725–740.
7. Florea, D. and S. Florea, *Big Data and the ethical implications of data privacy in higher education research*. Sustainability, 2020. 12(20): p. 8744.
8. Dhirani, L.L., Mukhtiar, N., Chowdhry, B.S. and Newe, T., Ethical dilemmas and privacy issues in emerging technologies: A review. *Sensors*, 2023. 23(3): p. 1151.
9. Ferretti, A. and E. Vayena, In the shadow of privacy: Overlooked ethical concerns in COVID-19 digital epidemiology. *Epidemics*, 2022. 41: p. 100652.
10. Heimstädt, M. and L. Dobusch, Transparency and accountability: Causal, critical and constructive perspectives. *Organization Theory*, 2020. 1(4): p. 2631787720964216.
11. Lindhout, P. and G. Reniers, The "transparency for safety" triangle: Developing a smart transparency framework to achieve a safety learning community. *International Journal of Environmental Research and Public Health*, 2022. 19(19): p. 12037.

12. Kashyap, S. and E. Iveroth, Transparency and accountability influences of regulation on risk control: The case of a Swedish bank. *Journal of Management and Governance*, 2021. 25(2): pp. 475–508.

13. Horstead, A. and A. Cree, Achieving transparency in forensic risk assessment: A multimodal approach. *Advances in Psychiatric Treatment*, 2013. 19(5): pp. 351–357.

14. Grundy, Q., Commentary–from transparency to accountability: Finding ways to make expert advice trustworthy. *Healthcare Policy*, 2022. 17(3): p. 28.

15. Duchek, S., S. Raetze, and I. Scheuch, The role of diversity in organizational resilience: A theoretical framework. *Business Research*, 2020. 13(2): pp. 387–423.

16. Liu, J., Y. Zhu, and H. Wang, Managing the negative impact of workforce diversity: The important roles of inclusive HRM and employee learning-oriented behaviors. *Frontiers in Psychology*, 2023. 14: p. 1117690.

17. Pandi-Perumal, S.R., Akhter, S., Zizi, F., Jean-Louis, G., Ramasubramanian, C., Edward Freeman, R. and Narasimhan, M., Project stakeholder management in the clinical research environment: How to do it right. *Frontiers in Psychiatry*, 2015. 6: p. 71.

18. Renn, O., Stakeholder and public involvement in risk governance. *International Journal of Disaster Risk Science*, 2015. 6: pp. 8–20.

19. Jeffery, N., Stakeholder engagement: A road map to meaningful engagement. *Doughty Centre, Cranfield School of Management*, 2009. 2: pp. 19–48.

20. Kaptein, M., A paradox of ethics: Why people in good organizations do bad things. *Journal of Business Ethics*, 2023. 184(1): pp. 297–316.

21. Stahl, B.C., Antoniou, J., Ryan, M., Macnish, K. and Jiya, T., Organisational responses to the ethical issues of artificial intelligence. *AI & Society,* 2022. 37(1): pp. 23–37.

22. Lindhout, P. and G. Reniers, Involving moral and ethical principles in safety management systems. *International Journal of Environmental Research and Public Health*, 2021. 18(16): p. 8511.

23. San-Jose, L., J.F. Gonzalo, and M. Ruiz-Roqueni, The management of moral hazard through the implementation of a Moral Compliance Model (MCM). *European Research on Management and Business Economics*, 2022. 28(1): p. 100182.

24. Samans, R. and J. Nelson, Corporate strategy and implementation, in *Sustainable enterprise value creation: Implementing stakeholder capitalism through full ESG integration*. 2022, Springer. pp. 141–186.

25. Cakir, M.S., J.K. Wardman, and A. Trautrims, Ethical leadership supports safety voice by increasing risk perception and reducing ethical ambiguity: Evidence from the COVID-19 pandemic. *Risk Analysis*, 2023. 43(9): pp. 1902–1916.

26. Settembre-Blundo, D., González-Sánchez, R., Medina-Salgado, S. and García-Muiña, F.E., Flexibility and resilience in corporate decision making: a new sustainability-based risk management system in uncertain times. *Global Journal of Flexible Systems Management*, 2021. 22(Suppl 2): pp. 107–132.

27. Hurst, S.A. and A. Mauron, A question of method: The ethics of managing conflicts of interest. *EMBO Reports*, 2008. 9(2): pp. 119–123.

28. Østengaard, L., Lundh, A., Tjørnhøj-Thomsen, T., Abdi, S., Gelle, M.H., Stewart, L.A., Boutron, I. and Hróbjartsson, A., Influence and management of conflicts of interest in randomised clinical trials: qualitative interview study. *BMJ*, 2020. 371: pp. 1–10.

29. Wiltshire, D. and S. Alvanides, Ensuring the ethical use of big data: Lessons from secure data access. *Heliyon*, 2022. 8(2): p. e08981.

30. Bormida, M.D., The big data world: Benefits, threats and ethical challenges, in *Ethical issues in covert, security and surveillance research*. 2021, Emerald Publishing Limited. pp. 71–91.

31. Struminskaya, B. and J.W. Sakshaug, Ethical considerations for augmenting surveys with auxiliary data sources. *Public Opinion Quarterly*, 2023. 87(S1): pp. 619–633.

32. Taherdoost, H., Data collection methods and tools for research; A step-by-step guide to choose data collection technique for academic and business research projects. *International Journal of Academic Research in Management (IJARM)*, 2021. 10(1): pp. 10–38.

33. Sanjari, M., Bahramnezhad, F., Fomani, F.K., Shoghi, M. and Cheraghi, M.A., Ethical challenges of researchers in qualitative studies: The necessity to develop a specific guideline. *Journal of Medical Ethics and History of Medicine*, 2014. 7: pp. 1–6.

34. Bennouna, C., H. Mansourian, and L. Stark, Ethical considerations for children's participation in data collection activities during humanitarian emergencies: A Delphi review. *Conflict and Health*, 2017. 11: pp. 1–15.

35. Boustead, A.E. and T. Herr, Analyzing the ethical implications of research using leaked data. *PS: Political Science & Politics*, 2020. 53(3): pp. 505–509.

36. Howse, R. and J. Langille, Continuity and change in the World Trade Organization: Pluralism past, present, and future. *American Journal of International Law*, 2023. 117(1): pp. 1–47.

37. Gleißner, W., T. Günther, and C. Walkshäusl, Financial sustainability: measurement and empirical evidence. *Journal of Business Economics*, 2022. 92(3): pp. 467–516.

38. Haworth-Brockman, M., C. Betker, and Y. Keynan, Saying it out loud: Explicit equity prompts for public health organization resilience. *Frontiers in Public Health*, 2023. 11: p. 1110300.

39. Aung, K., Rahman, N., Nurumal, M.S. and Ahayalimudin, N., Ethical disaster or natural disaster? Importance of ethical issues in disaster management. *Journal of Nursing and Health Science*, 2017. 6(2): pp. 90–93.

40. Kwak, Y., J.-W. Ahn, and Y.H. Seo, Influence of AI ethics awareness, attitude, anxiety, and self-efficacy on nursing students' behavioral intentions. *BMC Nursing*, 2022. 21(1): p. 267.

41. Alahmari, S., K. Renaud, and I. Omoronyia, Moving beyond cyber security awareness and training to engendering security knowledge sharing. *Information Systems and e-Business Management*, 2023. 21(1): pp. 123–158.

42. Macnish, K. and J. Van der Ham, Ethics in cybersecurity research and practice. *Technology in Society*, 2020. 63: p. 101382.

43. Roy, A., Newman, A., Round, H. and Bhattacharya, S., Ethical culture in organizations: A review and agenda for future research. *Business Ethics Quarterly*, 2024. 34(1): pp. 97–138.

44. Falkenberg, L. and I. Herremans, Ethical behaviours in organizations: Directed by the formal or informal systems? *Journal of Business Ethics*, 1995. 14: pp. 133–143.

12 Ethics in Information Technology in Canada

12.1 CANADIAN LEGAL AND REGULATORY FRAMEWORKS FOR IT ETHICS

Canada has proactively approached the creation of legislative and regulatory frameworks to guarantee the safe, ethical, and equitable application of artificial intelligence (AI) and other emerging technologies. The Canadian federal government is working together to offer a regulatory environment that is both friendly to innovation and intended to lower the possibility of possible risks [1].

A keystone of the initiatives being undertaken by Canada is the creation of the AI and Data Act (AIDA), which aims to offer a strict legal framework for the application of AI. In an attempt to build public confidence and promote the growth of Canadian AI companies, the government has also created an optional AI Code of Conduct and particular guidelines for applying generative AI in federal agencies. Canada is committed to creating a single national AI plan, as seen by the Pan-Canadian Artificial Intelligence initiative, which receives significant government funding [1].

Provincial variations hamper the coordination of privacy needs to be improved at the national level in personal (health) information protection laws and healthcare data governance plans. Conversely, federal and provincial privacy commissioners play a major role in supervising companies' security and privacy policies handling personal data [2]. Essential and legally required supervisors in data-intensive research are research ethics boards (REBs). This is particularly true when researchers wish to obtain personal data without the people's permission [2, 3].

In Canada, a network of rules at multiple levels provides a basis for clinical research even though no clear law controls the conduct of research involving human participants. Important suggestions include the International Council for Harmonization of Technical Requirements for Pharmaceuticals for Human Use—Guidance for Good Clinical Practice and the Tri-Council Policy Statement: Ethical Conduct for Research Involving Humans (TCPS2) [3]. Conversely, some argue that the somewhat uncontrolled nature of REB activities in Canada offers a chance to improve the current system's effectiveness [3, 4].

12.1.1 FEDERAL LEGISLATION GOVERNING IT ETHICS

Commercial entities in Canada are obligated by federal law to gather personal information only with the express permission of individuals or under certain conditions [5]. This legislative obligation emphasizes the need to uphold people's privacy rights and ensure data-collecting techniques are used honestly and morally. International norms and principles about protecting personal information and privacy were intended to be followed in establishing Canada's regulatory environment. By 2020, Canada will have committed to following international standards in information technology (IT) ethics when the G20 leaders ratify the OECD's guidelines [6]. By acknowledging and including these international agreements in its legal system, Canada is dedicated to upholding moral standards in the IT industry and guaranteeing compliance with larger multinational projects.

Technology has advanced, and with it, the introduction of AI and strong data analytics has made it imperative to reassess the current legislative frameworks to handle new challenges and ethical issues. Professional responsibilities and ethical requirements must be included in codes of ethics and professional conduct to direct ethical behavior within the IT sector successfully. Protection of personal information, preventing harmful factors, and advancing human well-being are a few instances [6]. The fact that these principles exist provides a foundation for moral judgments and emphasizes the need to consider moral issues while developing new technology.

12.1.2 PROVINCIAL LEGISLATION AND JURISDICTIONAL VARIANCES

Each province has its legislative authority. Therefore, IT ethics policies and laws may vary. Provincial legislation may establish protections or obligations particular to the province's goals. Provincial privacy and technology legislation may apply to education and healthcare. Businesses with operations in different provinces may need help with jurisdictional differences. They must traverse a maze of restrictions to comply with provincial and federal laws. This may require changing organizational policies and practices to meet each jurisdiction's requirements while maintaining internal coherence and harmony.

IT ethics enforcement may be affected by jurisdictional resource and priority shortages. National compliance and enforcement outcomes vary because provinces dedicate different enforcement resources or prioritize various ethical issues. Organizations must proactively educate themselves on new federal and provincial regulations, stay abreast of the latest developments, and establish robust compliance systems that account for jurisdictional variations to comply in Canada. Collaboration with trade associations and legal specialists can help organizations follow ethical standards and understand provincial laws.

12.1.3 INDUSTRY STANDARDS AND CODES OF CONDUCT

Standards and codes of conduct within the Canadian IT industry are crucial in establishing what is considered to be the most ethical way to do business. Businesses and IT experts can look to this collection of regulations and standards as a road map. The norms of conduct, decision-making, and interactions are laid out in them.

Industry-specific associations and organizations often create and update these standards to promote consistency and responsibility across the board. The Canadian Advanced Technology Alliance (CATA) and the Canadian Information Processing Society (CIPS) are prominent examples of organizations that help shape IT ethics.

In addition to being a matter of morality, the Canadian IT industry's credibility and reliability are enhanced when members adhere to these standards and codes of behavior. These standards are essential for the industry's reputation and integrity because they encourage a culture of honesty and responsibility. These standards and codes typically address various aspects of ethical conduct, as shown in Figure 12.1.

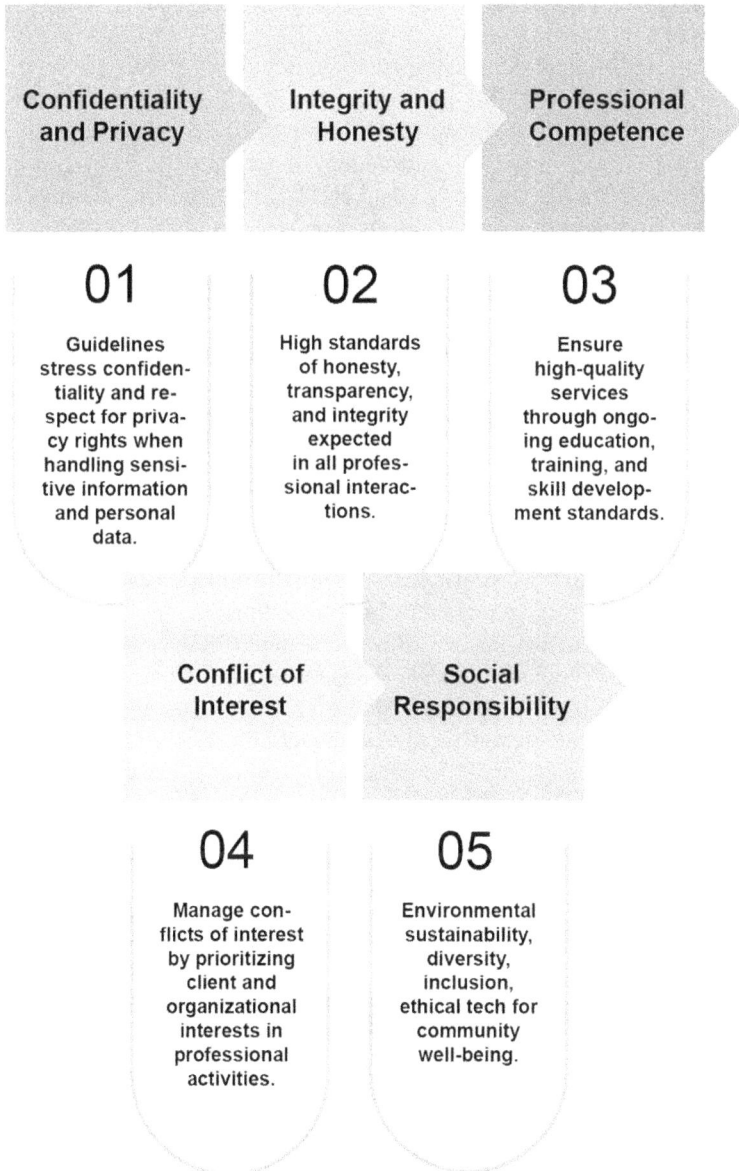

Confidentiality and Privacy

01

Guidelines stress confidentiality and respect for privacy rights when handling sensitive information and personal data.

Integrity and Honesty

02

High standards of honesty, transparency, and integrity expected in all professional interactions.

Professional Competence

03

Ensure high-quality services through ongoing education, training, and skill development standards.

Conflict of Interest

04

Manage conflicts of interest by prioritizing client and organizational interests in professional activities.

Social Responsibility

05

Environmental sustainability, diversity, inclusion, ethical tech for community well-being.

FIGURE 12.1 Ethical guidelines for professional conduct

12.1.4   REGULATORY BODIES OVERSEEING IT ETHICS COMPLIANCE

Under the Personal Information Protection and Electronic Documents Act (PIPEDA), the federal body responsible for looking into privacy complaints and negotiating compliance agreements with businesses is the Office of the Privacy Commissioner of Canada (OPC) [2]. Meanwhile, Proposals have been made to strengthen the OPC's jurisdiction to maintain compliance with the European Union's General Data Protection Regulation (GDPR). Now, acting as an ombudsperson, the OPC cannot issue orders or levy fines.

Several privacy commissioners possess more powerful enforcement instruments at the provincial level, including the capacity to fine. Researchers accessing data from Europe should know the steep penalties for non-compliance with the GDPR and its possible extraterritorial implementation. Importantly, REBs in Canada oversee research that primarily depends on data. REBs are empowered to waive permission requirements where identifiable information is required for the study, adequate privacy protections are in place, and persons have not voiced objections. Accessing healthcare data in the Quebec province may require researchers to get further permission from administrative bodies and organizations [2].

Clinical research in Canada is regulated by the Food and Drug Regulations, which are a part of the Food and Drugs Act, and the Tri-Council Policy Statement: Ethical Conduct for Research Involving Humans (TCPS2), which mandates that organizations receiving public funding from the Natural Sciences and Engineering Research Council, the Social Sciences and Humanities Research Council, or the Canadian Institutes of Health Research must adhere to ethics review compliance standards. The Department of Health and Human Sciences sponsors research conducted in Canada. Hence, compliance with U.S. regulations is essentially imported even though Canada lacks a consistent government research ethics regulation [3].

The Interagency Advisory Panel on Research Ethics (PRE) was established by the three federal research bodies in Canada and is tasked with offering unbiased advice on the morality of human research [2]. Moreover, the Canadian General Standards Board and Health Canada have developed a standard for the research ethics supervision of health research in Canada [3]. Table 12.1 provides a succinct synopsis of the key regulatory bodies in Canada, outlining their areas of responsibility and main foci in ensuring IT ethical compliance.

12.1.5   INTERNATIONAL AGREEMENTS AND THEIR IMPACT ON CANADIAN IT ETHICS

Canadian IT research and development ethics can be affected by international agreements. These agreements often set ethical standards for IT professionals and researchers. The UN Declaration on the Rights of Indigenous Peoples (UNDRIP) guarantees Indigenous peoples the right to maintain, control, safeguard, and develop their cultural heritage, traditional knowledge, and traditional cultural expressions. In Canada, community-based Indigenous research ethical guidelines like OCAP® (Ownership, Control, Access, and Possession) protect persons and communities' privacy [7]. These procedures ensure that Indigenous community research is ethical and respects their rights.

EU GDPR is also relevant. While the GDPR does not apply to Canada, it has influenced Canadian data privacy laws and practices. Personal data collection, use, and transfer must comply with the GDPR, including explicit consent. Canada has developed standardized data transfer and use agreements (DTUAs) to secure personal data when shared between organizations. Other worldwide frameworks for research ethics include the World Medical Association's Declaration of Helsinki and the Council for Worldwide Organizations of Medical Sciences (CIOMS) International Ethical Guidelines for Human Health Research [5].

TABLE 12.1
Overview of regulatory bodies in Canadian IT ethics compliance

Regulatory Body	Responsibilities	Jurisdiction	Primary Focus
Office of the Privacy Commissioner of Canada (OPC)	Oversees privacy compliance and investigates complaints.	Federal	Privacy and Data Protection
Canadian Radio-television and Telecommunications Commission (CRTC)	Regulates broadcasting and telecom, enforces laws.	Federal	Broadcasting and Telecommunications
Canadian Cyber Incident Response Centre (CCIRC)	Coordinates national response to cyber threats.	Federal	Cybersecurity Incident Response
Canadian Standards Association (CSA)	Develops industry standards and guidelines.	Non-governmental	Industry Standards and Certification
Information and Privacy Commissioner Offices at Provincial/Territorial Levels	Enforces provincial privacy laws.	Provincial/ Territorial	Privacy and Data Protection Laws Enforcement

12.2 ETHICAL CHALLENGES AND CONSIDERATIONS IN CANADIAN IT INDUSTRY

Increasingly, Canadian companies are using AI. Among the many applications of AI in these companies are data analysis, fraud detection, and CRM analysis. Business leaders need help to address ethical issues even if they understand their importance; a few believe ethical issues seriously jeopardize their AI projects. This shows that the significance of ethical AI practices being acknowledged and their practical application in the Canadian business sector needs to be aligned.

Research on digital communication ethics demonstrates that ethical issues are not exclusive to AI applications. Comparing American professionals to their Canadian counterparts, the former reported more ethical problems, and a larger proportion had ethics training. As such, managing the complexities of digital communication in business settings requires training and ethical standards [8]. A comprehensive strategy is necessary since ethical issues in the always-evolving digital ecosystem are distributed unevenly among demographics and geographic regions.

Regarding new technology, especially quantum technologies, there are grave moral problems. Though politics, society, the law, and ethics are all hotly debated, quantum technology has a lot of possible advantages. A responsible adoption of quantum technologies requires careful consideration of data security, privacy, regulatory issues, and fair access to technology.

12.2.1 DATA PRIVACY AND SECURITY CONCERNS

Data privacy concerns encompass all corporate acquisition, storage, and use of personal information. Citizens of Canada are entitled to control how their data is gathered, used, and disclosed. Still, as social media, e-commerce, and online services have grown, customers often need help understanding and managing their digital footprints.

A key ethical issue is still data security, especially given cyber threats' growing complexity and reach. Cyberattacks and data breaches can cause substantial financial losses, harm one's reputation, and violate someone's right to privacy. As such, companies operating in Canada must implement strict security protocols to guard against unwanted access, data breaches, and other types of cyberattacks. New complexities to the data privacy and security field have been introduced by the development of technologies like AI, big data analytics, and the Internet of Things (IoT). Even if new technologies offer revolutionary opportunities for efficiency and creativity, they also raise moral questions about data's appropriate and moral use. Among the issues brought up are algorithmic unfairness, discriminatory effects, and the degradation of privacy rights brought about by pervasive surveillance and data profiling.

The Canadian regulatory bodies created laws like the PIPEDA to establish guidelines for private sector companies to gather, process, and distribute personal data. Furthermore, privacy commissioners at the federal and provincial levels are crucial to ensuring data protection laws are obeyed and looking into complaints linked to privacy infractions.

12.2.2 ACCESSIBILITY AND INCLUSIVITY IN TECHNOLOGICAL DEVELOPMENT

A legislative framework has been established to encourage inclusive and accessible technology. One notable example is the Accessibility for Ontarians with Disabilities Act (AODA), which mandates accessibility standards for Ontario organizations, whether the public or private sector. Disabled people should be able to access and utilize technology with the same ease as everyone else; the Canadian Human Rights Act forbids discrimination based on disability.

While new technologies could drastically improve accessibility for people with impairments, they also present obstacles that need to be considered before implementation. Websites and apps that do not work with screen readers or do not have keyboard shortcuts can make it very difficult for those with motor impairments or blindness to use the internet.

While creating a product, developers and designers should keep accessibility and inclusion in mind. Web content accessibility guidelines (WCAG) implementation, user testing with people of varying abilities, and alternative content formats (e.g., transcripts of audio recordings and closed captions for videos) are all part of this effort.

The tech industry must promote an inclusive culture to make a real difference. As part of this effort, we must aggressively seek the opinions of underrepresented groups throughout development, foster inclusive workplaces, and encourage diversity in recruiting procedures. Technology developed in Canada can better serve a varied population and promote equality and social justice if its developers emphasize accessibility and inclusivity.

12.2.3 ENVIRONMENTAL IMPACT OF IT INFRASTRUCTURE

From energy usage to electronic waste management, IT infrastructure's fast expansion has major environmental consequences. Recognizing and reducing the environmental impact is paramount as Canada's IT sector grows. The high energy consumption of IT infrastructure is a major cause for concern. Servers and networking gear in data centers necessitate massive power for cooling and powering the facility. The increase in greenhouse gas emissions caused by this energy demand worsens climate change. Additionally, the proliferation of devices like computers, cellphones, and IoT gadgets increases the total energy consumption of IT systems.

Energy efficiency measures for IT infrastructure are now being implemented. This effort includes improved hardware design, better data center cooling systems, and the use of renewable energy sources to power IT operations. Less physical hardware is required, and energy consumption is decreased, thanks to virtualization and cloud computing technologies, which also allow for more efficient use of resources.

The disposal of electronic equipment is becoming an increasingly serious environmental concern due to the rapid rate at which technology is evolving and devices are becoming outdated. Communities exposed to harmful substances from improperly handled e-waste face health risks and environmental contamination.

Canadian legislation and recycling initiatives encourage the proper disposal and recycling of electronic gadgets to lessen the negative effects of this trash on the environment. By promoting the reuse and recycling of valuable components contained in electronic equipment, these projects hope to divert e-waste from landfills.

Many IT products and infrastructure are being developed with sustainable design concepts. Reduced waste and resource depletion during a product's lifetime can be achieved by using eco-friendly materials, long-lasting and repairable design, and implementing circular economy models.

12.2.4 INTELLECTUAL PROPERTY RIGHTS AND FAIR USE

Considerations of fair use and intellectual property (IP) rights have a major impact on how the Canadian IT industry fosters innovation and creativity. The rules and regulations that control these rights aim to strike a balance between the public interest and private ownership while preserving the interests of innovators, creators, and companies.

Trade secrets, trademarks, copyrights, and patents are Canada's main components of IP rights. Software code, databases, and digital content are all considered works of literary, artistic, musical, or dramatic creations that their creators exclusively own according to copyright laws. A patent grants the owner the legal right to make, use, and sell their patented product or procedure for some time. Trade secrets protect secret information that gives companies an edge in the market, while trademarks protect the names, logos, and symbols that identify products and services.

Research, private study, criticism, review, news reporting, education, parody, and satire are all examples of uses that can be exempted from copyright rules under the Canadian concept of fair use or fair dealing. Acknowledgment of the source and fair use of the work are two conditions that must be met for fair dealing to occur. The Canadian IT industry faces ethical challenges and considerations related to IP rights and fair use, as shown in Figure 12.2.

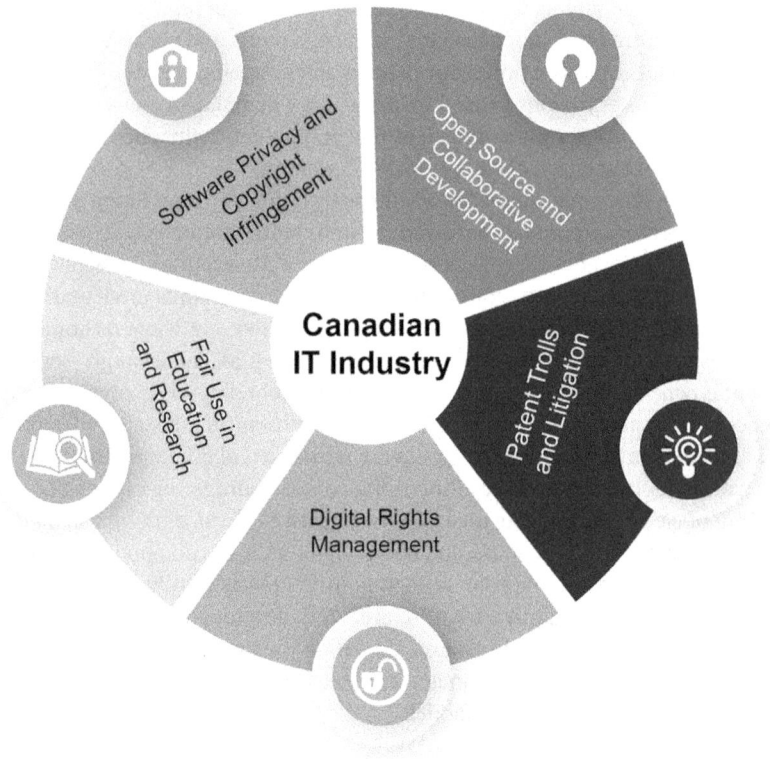

FIGURE 12.2   Canadian IT industry ethics issues

12.3   PRIVACY AND DATA PROTECTION LAWS IN CANADA

12.3.1   Personal Information Protection and Electronic Documents Act (PIPEDA)

An important Canadian federal law controlling private entities' gathering, use, and disclosure of personal information is the PIPEDA. For businesses operating in jurisdictions without privacy laws that are substantially equivalent to PIPEDA, the 2000 law applies. It also regulates storing and disclosing personally identifiable information in cross-border business dealings.

When collecting, using, or disclosing personal information, organizations must get an individual's consent under PIPEDA. In order for this permission to be valid, individuals need to know exactly what data is being gathered, how it will be utilized, and to whom it will be shared. Additionally, businesses must take reasonable precautions to prevent customer data loss, abuse, or alteration.

A person's right to access personal information stored by an organization and ask for errors to be rectified are both outlined in PIPEDA. Additionally, it requires enterprises to be open and honest about their privacy policies and procedures and to name specific people to ensure compliance.

The Privacy Commissioner of Canada is appointed by PIPEDA to ensure that the law is followed and to look into complaints about privacy violations. Orders, compliance agreements, and public reports on investigations and findings are all within the Commissioner's purview.

As new privacy concerns and technologies have arisen, PIPEDA has evolved to incorporate new sections, such as those about electronic documents and online activity. Canada is known for its strong privacy safeguards, which is largely due to the fact that its values align with those of other countries. Nevertheless, there are ongoing discussions over whether PIPEDA is sufficient to address the changing nature of privacy threats and whether it needs to be updated to protect personal information effectively in the digital era.

12.3.2   Cross-Border Data Transfers and Compliance with Privacy Laws

The interconnectivity of the digital world has forced Canadian businesses to send data abroad. There are concerns, meanwhile, about upholding privacy laws and safeguarding individual rights during such transfers.

In Canada, private sector companies who gather, use, or disclose personal information for commercial reasons are subject to PIPEDA. With rare exceptions, PIPEDA mandates that Canadian personal data be exported with meaningful consent.

Before sending data abroad, Canadian firms must confirm that the target country offers at least as strong privacy protections for personal information as Canada. To assess a measure's appropriateness, consider the legal framework, enforcement mechanisms, and protections for the personal data of the target nation.

Cross-border data transfers can be facilitated while respecting privacy regulations using standard contractual clauses (SCCs) approved by regulatory bodies. Legal responsibilities to safeguard personal data during transmission are placed on sending and receiving parties under these rules.

Binding corporate regulations (BCRs) is another way Canadian multinational corporations might disclose personal information outside contractual channels. BCRs specify internal data security guidelines.

Cross-border data transfers must be watched, and Canadian privacy authorities must guarantee privacy compliance. Companies sending personal data abroad have to protect privacy rights fairly. This covers identifying risks and exercising safety measures.

12.4   ADDRESSING ETHICAL IMPLICATIONS OF CANADIAN CYBERSECURITY STRATEGIES

12.4.1   CYBERSECURITY THREAT LANDSCAPE IN CANADA

Threats to Canada's cybersecurity are complex and ever-changing, as they are for many other industrialized nations. A wide range of cyber threats, including those from nation-states, criminal organizations, hacktivists, and malevolent insiders, are becoming more prevalent as the nation's essential infrastructure, public services, and private sector operations become more digitalized at a rapid pace.

The widespread nature of cybercrime poses a serious threat to Canada's cybersecurity infrastructure. Businesses, government institutions, and individuals are all targets of criminal groups' data breaches, financial fraud, identity theft, and ransomware assaults. In addition to draining bank accounts, these assaults destroy public faith in digital processes and cause people to lose faith in government altogether.

The Canadian government is also frequently the object of cyber assaults and espionage by foreign powers. The nation's crucial infrastructure, substantial IP, and membership in NATO make it an attractive target for political and economic opponents. Attacks on critical infrastructure, theft of private data, and threats to national security are all goals of state-sponsored cyber operations.

Foreign cyber threats can have far-reaching consequences for Canadian organizations and individuals due to the interconnectedness of the global digital economy. Law enforcement and cybersecurity organizations face difficulties recognizing, attributing, and mitigating cross-border cyber attacks orchestrated by foreign adversaries or transnational criminal networks.

The Canadian government has responded to these threats by increasing spending on cybersecurity and working to improve cooperation with foreign allies. Efforts are being made by organizations like the Cyber Security Cooperation Program (CSCP), the National Cyber Threat Assessment Centre (NCTAC), and the Canadian Centre for Cyber Security (CCCS) to raise awareness about cybersecurity, coordinate responses to incidents, and share threat intelligence.

Regardless of these endeavors, the cybersecurity landscape in Canada is ever-changing and demands constant attention and adjustment. Organizations and individuals in Canada must prioritize cybersecurity measures like proactive threat detection, employee training, and robust risk management to protect digital assets and interests from cyber threats and keep up with the ever-changing technology landscape.

12.4.2   ETHICAL CONSIDERATIONS IN CYBER DEFENSE AND OFFENSE

When private companies engage in defensive actions and more controversial tactics like active-cyber defense and "hacking back," it raises serious ethical concerns within commercial cybersecurity operations. Similar ethical concerns exist with using private military and security corporations (PMSCs) for cybersecurity services as with using private businesses for traditional military and security services. To address these ethical concerns, this situation calls into question the level of involvement of the private sector in cyber defense and offense, highlighting the necessity of a moderately stringent strategy [9].

In the field of cybersecurity, ethical hacking is essential, as it promotes the use of unauthorized access to systems in order to strengthen cyber resilience. Recognizing that the question is not if but when a cyberattack will happen, organizations must take proactive measures to prepare for cyber risks. In order to reduce the likelihood of harm and maintain ethical standards in cybersecurity, it is crucial to have strategies in place for emergency responses. Some ethical difficulties raised by cybersecurity include obtaining employees' informed consent before monitoring them and the moral dilemmas that arise when cybersecurity risks are reduced at the price of customers' privacy [10].

The ever-changing digital ecosystem presents cybersecurity experts and companies in Canada and many other countries with ethical concerns and privacy issues. Justice, nonmaleficence, explicability, beneficence, and autonomy comprise cyber ethics' foundation and give a framework for making ethical decisions in cybersecurity activities. Upholding ethical standards while negotiating the intricacies of cybersecurity operations in Canada and worldwide is crucial in addressing these ethical problems, necessitating a thorough grasp of cyber defense and offense [11].

12.4.3 Public-Private Collaboration in Cybersecurity

Canada is studying the pros and cons of public-private partnerships (PPPs) in cybersecurity as global adoption grows. Governmental-private partnerships (GPPs) in cybersecurity governance aim to combine governmental and private sector strengths. This recognizes that cyber threats demand a coordinated strategy. Canada is embracing a collaborative and flexible approach to cybersecurity by encouraging information sharing, offering incentives for cooperation, and building a strong security framework to protect vital infrastructure. This focused endeavor strengthens the country's cyber defenses and sets the bar for global cybersecurity collaboration by creating new laws and mechanisms that foster cross-sector cooperation [12]. Figure 12.3 shows how public and private cybersecurity sectors share resources, experience, and information.

Public and private sectors can collaborate on cybersecurity through information sharing, cooperation, and partnerships. The non-profit Canadian Cyber Threat Exchange (CCTX) fosters information exchange and teamwork to increase cybersecurity resilience in all industries.

Information sharing lets public and private companies share threat intelligence, compromise signs, and best practices. Stakeholders can increase their defenses and response capabilities by being informed about new vulnerabilities and threats. Coordination and cooperative projects combine resources, knowledge, and technology to build creative cybersecurity plans and solutions.

Public-private partnerships often include government funding, regulatory frameworks, and other cybersecurity assistance. Cybersecurity regulations and standards, financial incentives for cybersecurity efforts, and industry-wide coordination are feasible tactics. Business partners provide expertise, skills, and resources to develop the nation's cybersecurity.

In public-private cooperation, balancing the need to share information with concerns about privacy, secrecy, and economic advantage can be difficult. Private data and participant confidence must be protected to promote efficient cooperation arrangements. Diversity and inclusion programs in cooperative teams may inspire innovative and effective cybersecurity processes.

FIGURE 12.3   Cybersecurity partnerships between public and private sectors

12.4.4 CYBERSECURITY EDUCATION AND AWARENESS PROGRAMS

These initiatives educate consumers on cyber threats, digital asset protection, and security challenges. Canada has many public, corporate, and academic cybersecurity awareness and education programs. Public Safety Canada and the CCCS collaborate with academic institutions, charities, and businesses to develop and implement cybersecurity awareness programs. These initiatives use social media, workshops, seminars, and internet resources to educate people about phishing, malware, and ransomware.

Canada's educational institutions offer cybersecurity courses, training, and certifications to train the next generation of cybersecurity experts. These courses include ethical hacking, network security, cryptography, and incident response. These educational activities build a resilient cybersecurity workforce that can mitigate growing cyber threats by providing the necessary information and skills.

Community-based groups and grassroots initiatives help Canadians understand cybersecurity beyond official instruction. These projects aim to engage elders, small businesses, and Indigenous communities, who may be more vulnerable to cyber-attacks due to resource or awareness gaps. These initiatives focus on outreach to these communities' needs and challenges to provide individuals with internet safety knowledge and services.

12.5   INDIGENOUS RIGHTS AND ETHICAL USE OF TECHNOLOGY IN CANADA

Indigenous peoples' perspectives on technology and ethics are based on cultural traditions, spiritual beliefs, and a holistic worldview emphasizing connectivity with the natural world. For many Indigenous communities in Canada, technology is more than just a tool for environmental stewardship, cultural preservation, and community well-being. It has significant ethical ramifications.

Indigenous tribes typically see technology through the lenses of sustainability and respect for the land. They recognize the need to balance human activity and the environment. This perspective challenges conventional notions of development and advancement by prioritizing the long-term health of ecosystems over short-term gains.

Indigenous cultures place a higher priority on community decision-making and communal values than they do on technology as a standalone goal. Technology is evaluated not just for its usefulness but also for how well it conforms to existing knowledge and how it may impact aspects such as cultural preservation and societal harmony.

Regarding ethical considerations surrounding technology in Indigenous contexts, questions like cultural autonomy, self-determination, and the freedom to regulate the use of traditional knowledge loom larger than issues of privacy and data sovereignty. Indigenous people have long expressed concern that technologies such as digital media, biotechnology, and IP laws will result in the exploitation and commercialization of their traditional knowledge.

As they negotiate the intersection of technology and ethics, Indigenous communities are battling for methods to innovate that are culturally appropriate, strengthen local governance structures, and foster respectful partnerships with external stakeholders. This broader movement includes initiatives to bridge the digital divide, support Indigenous-led research and development, and promote integrating traditional ecological knowledge into contemporary solutions.

An ethical factor in Indigenous data sovereignty is respecting the rights of Indigenous communities to manage and control their data in line with their cultural values, political structures, and right to self-determination. Given the possibility of the inclusion of sensitive spiritual, cultural, and traditional knowledge, it is imperative to recognize the unique significance of Indigenous data. Ethical norms demand that consent be obtained freely, prior to, and informedly before collecting, using, or disclosing data from Indigenous people. Furthermore, it is critical to ensure that the use of data benefits Indigenous peoples and respects their autonomy, dignity, and right to privacy. Establishing data ownership, governance, and stewardship mechanisms can empower Indigenous communities and safeguard their data against exploitation, misuse, or theft.

In order to achieve the ethical and reconciliation obligations, Canada has to approach the development and use of technology with care and respect for Indigenous people. In order to accomplish this, we must first accept the wrongs of the past, pay attention to the opinions of Indigenous peoples regarding technology, and involve them in choices about its usage in their culture and on their territory. Above all, it is about ensuring that ICT initiatives support Indigenous peoples rather than escalating problems or reinforcing colonial legacies. It is imperative to adopt the ideas of consultation, collaboration, and permission to advance genuine reconciliation and uphold ethical duties toward Indigenous communities in the IT industry.

REFERENCES

1. Walter, Y., Managing the race to the moon: Global policy and governance in Artificial Intelligence regulation—A contemporary overview and an analysis of socioeconomic consequences. *Discover Artificial Intelligence*, 2024. 4(1): p. 14.
2. Thorogood, A., Canada: Will privacy rules continue to favour open science? *Human Genetics*, 2018. 137(8): pp. 595–602.
3. Alas, J.K., Godlovitch, G., Mohan, C.M., Jelinski, S.A. and Khan, A.A., Regulatory framework for conducting clinical research in Canada. *Canadian Journal of Neurological Sciences*, 2017. 44(5): pp. 469–474.
4. Nicholls, S.G., Morin, K., Evans, L. and Longstaff, H., Call for a pan-Canadian approach to ethics review in Canada. *Cmaj*, 2018. 190(18): pp. E553–E555.
5. Scheibner, J., Ienca, M., Kechagia, S., Troncoso-Pastoriza, J.R., Raisaro, J.L., Hubaux, J.P., Fellay, J. and Vayena, E., Data protection and ethics requirements for multisite research with health data: A comparative examination of legislative governance frameworks and the role of data protection technologies. *Journal of Law and the Biosciences*, 2020. 7(1): p. lsaa010.
6. Kisselburgh, L. and J. Beever, The ethics of privacy in research and design: Principles, practices, and potential, in *Modern socio-technical perspectives on privacy*. 2022, Springer International Publishing Cham. pp. 395–426.
7. Hayward, A., Sjoblom, E., Sinclair, S. and Cidro, J., A new era of Indigenous research: Community-based Indigenous research ethics protocols in Canada. *Journal of Empirical Research on Human Research Ethics*, 2021. 16(4): pp. 403–417.
8. Meng, J., S. Kim, and B. Reber, Ethical challenges in an evolving digital communication era: Coping resources and ethics trainings in corporate communications. *Corporate Communications: An International Journal*, 2022. 27(3): pp. 581–594.
9. Pattison, J., From defence to offence: The ethics of private cybersecurity. *European Journal of International Security*, 2020. 5(2): pp. 233–254.
10. Morgan, G., *Ethical Issues in cybersecurity: Employing red teams, responding to ransomware attacks and attempting botnet takedowns*. 2021: Dublin City University.
11. Dhirani, L.L., Mukhtiar, N., Chowdhry, B.S. and Newe, T., Ethical dilemmas and privacy issues in emerging technologies: A review. *Sensors*, 2023. 23(3): p. 1151.
12. Kalisz A., *Public–Private Partnerships on Cybersecurity and International Law: Finding Multilateral Solutions*, in Ishikawa T, and Kryvoi Y (eds) Public and Private Governance of Cybersecurity: Challenges and Potential. *Cambridge University Press*, 2023: pp. 211–239.

13 AI Ethics in IT
From Design to Practical Implementation

13.1 FUNDAMENTALS OF AI

While the development of artificial intelligence (AI) can be traced back to millennia, its formal inception occurred during the 20th century with the 1956 Dartmouth Conference, where the term "AI" was first introduced (Figure 13.1). There was an initial push toward symbolic AI, but then there was an "AI winter" of disappointingly slow development and inflated expectations. However, expert systems and neural networks returned in the 1980s, paving the way for modern machine learning and deep learning advances. Virtual assistants, driverless vehicles, and other AI-powered technologies are proliferating today, but questions of prejudice and responsibility in AI ethics continue to take center stage.

An AI system must have several critical parts for it to work. Substantiating AI systems, software, and components with quality models is an important first step. Studies have shown that complete models are necessary for evaluating and improving the performance of AI software, systems, and components. It is crucial to comprehend and handle AI software, system, and component quality [1]. It is essential to integrate several AI methodologies when developing AI systems. Intelligent and clever systems that can handle real-world difficulties are created using machine learning, deep learning, advanced analytics, knowledge discovery, reasoning, and searching. To tailor and enhance AI systems for particular problem domains and applications, each of these methods adds something special to AI models' capabilities. Combined, these methods constitute AI-based modeling, enabling the intelligence and automation needed by contemporary systems [2].

If we want to make the most of AI, we need to know what kinds of AI there are. There are various AI models, each designed to solve problems somewhat differently. These models include analytical, functional, interactive, textual, and visual AI. Researchers and developers can tackle a wide range of problems in today's linked world by investigating and using these varied types of AI, leveraging the capabilities of each model [2]. With so many kinds of AI, the subject is ripe with opportunities for new developments and advancements.

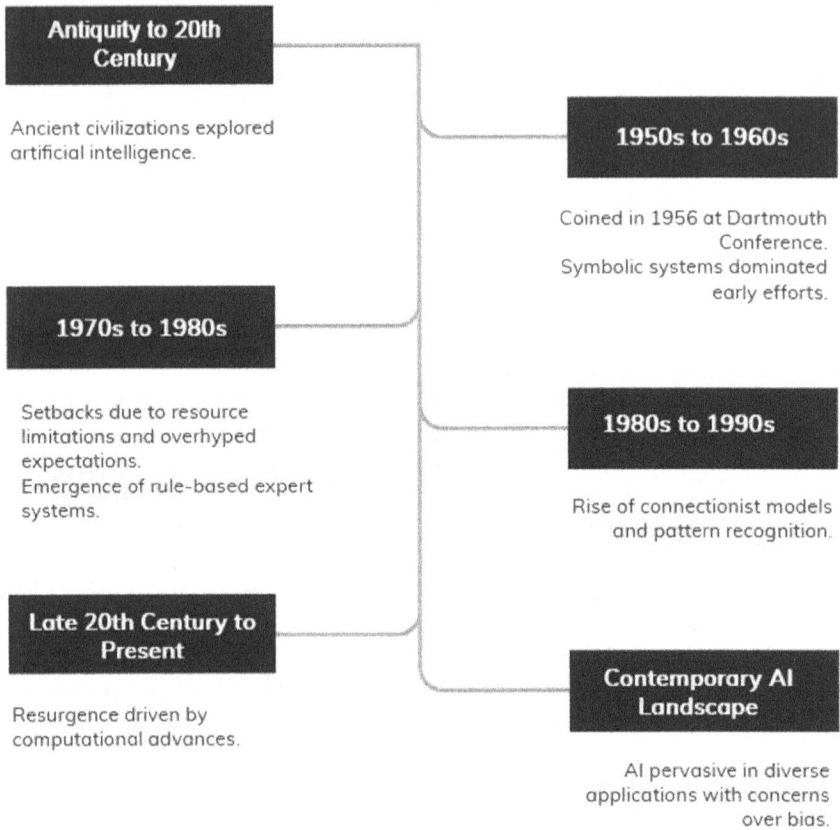

FIGURE 13.1 Evolution of AI

13.2 IMPLEMENTATION FRAMEWORKS FOR AI SYSTEMS

Many AI development projects have used Agile techniques in recent years. These methodologies allow teams to better adapt to changing needs, increase transparency, and promote continuous improvement [3–5]. The Agile methodology fits AI development's experimental and exploratory style due to its iterative and collaborative nature [3, 6]. With Agile, AI projects can reap several rewards, including:

- Adaptability to evolving needs and fresh perspectives from data and models [3, 4].
- Enhanced openness and cooperation among data scientists, subject matter experts, and stakeholders [3, 5].
- Quick iterative cycles lead to faster delivery of functional AI systems [4, 5].
- Based on user feedback and real-world performance, models are continuously learned and improved [3, 6].

The necessity for specific technical abilities, the management of AI project uncertainty, and the assurance of model interpretability and explainability are some of the unique issues of adapting Agile to AI development. To overcome these obstacles, researchers have put forward frameworks and methodologies, such as "data canyons," that would allow medical professionals to enhance AI models iteratively and agilely [6].

Efficiently managing the lifecycle of AI models now requires applying DevOps principles, like continuous automation. The deployment of AI solutions can be made faster, more reliable, and of higher quality if enterprises automate procedures like continuous delivery and monitoring [7]. The optimization of software delivery and operational efficiency in AI projects can be achieved with this technique, which enables the speedy and reliable deployment of numerous AI versions.

With the successful automation capabilities of DevOps extended to the deployment of AI models, the idea of AIOps or MLOps has arisen. By combining containers with microservices, AIOps builds data pipelines and AI models that are ready to be implemented. This modification uses containers to package apps, code, and dependencies to facilitate technology-agnostic hyper-automation for effective AI deployment. The data pipeline primarily deals with availability, preprocessing, and ethical considerations. In contrast, the AI pipeline ensures the scalability and dependability of AI solutions through model compression, maintenance, and monitoring [8].

A complete pipeline, including steps like Data Handling, Model Learning, Software Development, and System Operations, is necessary to develop AI models. Using DevOps principles as a roadmap, this pipeline simplifies AI solution lifecycle management from beginning to finish. Mapped to these stages are challenges connected to pipeline implementation, adaption, and usage; to overcome these challenges and ensure the effective deployment of AI systems, it is important to include DevOps approaches. Organizations may improve AI solution delivery's efficacy, dependability, and scalability by adopting a methodical strategy that aligns with DevOps principles [9]. By following these DevOps principles, companies may streamline AI system deployment, decrease manual work, and assure AI application reliability and scalability [10]. Key DevOps techniques for dependable and effective software development and deployment are listed in Table 13.1.

Pipelines for Continuous Integration (CI) and Continuous Deployment (CD) are fundamental to contemporary software creation methods. CI is a method for maintaining fully functional and up-to-date software by automatically integrating code changes into a shared repository daily. Software development can be more efficiently and dependably released with the help of CD, which builds upon CI by automating the deployment of code changes to production settings. The development process is made easier with these pipelines, which automate testing and deployment procedures, improve team collaboration, and decrease the risk of errors [12].

Software engineering relies on continuous integration and continuous delivery (CI/CD) pipelines to guarantee efficient and high-quality software development. With these pipelines, developers can automate the process of building, testing, and deploying code changes. This allows for consistent and rapid updates. In addition to shortening the development cycle, this automation improves software release dependability by finding defects early on. CI/CD pipelines encourage feedback and continual improvement, increasing development teams' agility and creativity [13].

Numerous technologies and solutions enable the automation of different phases of the software development lifecycle, which in turn supports the establishment of CI and CD pipelines. Several components, including automation servers, orchestrators, code retrievers, unit testers, artifact builders, and others, collaborate to form a streamlined pipeline that guarantees quick deployment of code changes after comprehensive testing. By automating the tedious but necessary processes involved in software development, testing, and deployment, these technologies free up developers to concentrate on feature implementation and code writing [14].

TABLE 13.1

Key DevOps practices for AI

Practice	Key Points	References
Continuous Integration (CI) and Continuous Deployment (CD)	Automates build, test, and deployment for frequent, reliable updates.	[7, 11]
Infrastructure as Code (IaC)	Programmatic provisioning and management for consistent deployment environments.	[11]
Monitoring and Logging	Tracks performance, data drift, and logs errors to ensure reliability.	[11]
Model Versioning and Management	Central repository for tracking changes and ensuring consistency.	[7]
Automated Testing	Automates various tests to catch issues early in development.	[9]
Containerization and Orchestration	Uses containers and Kubernetes for easy deployment, scaling, and resource management.	[7]

13.3   ETHICAL CONSIDERATIONS IN AI MODEL TRAINING

To guarantee equity, openness, and responsibility, the ethical considerations surrounding the training of AI models cover a wide range of important domains. In order to avoid the continuation of healthcare disparities and inequities, eliminating biases inside AI systems is a key ethical problem [15].

Responsible and compassionate AI research, especially about Large Language Models (LLMs) [15], requires scientific community-wide ethical discussions. Researchers and practitioners can use these conversations as a foundation to address potential ethical challenges during AI model training and develop frameworks for making ethical decisions. The scientific community may successfully traverse the complicated ethical landscape of AI research by participating in such discussions, which will guarantee that technological progress is in line with ethical principles and societal values.

Beyond the technical elements, there are broader societal consequences regarding the ethical aspects of AI model training. This all-encompassing method highlights the importance of ethical standards and frameworks that manage AI technology's social effects and ramifications in addition to the technical parts of training AI models [16]. Stakeholders can strive for a more just, equitable, and accountable AI ecosystem that respects basic human rights by including ethical concepts in AI model training procedures.

13.3.1   DATASET BIAS AND FAIRNESS

A dataset bias occurs when the training dataset contains skewed or unrepresentative data, which can cause AI models to produce biased results or predictions. Historical inequities, sample techniques, or data-gathering procedures are all potential origins of this bias. In order to combat dataset bias, it is crucial to use training data that is both diverse and representative to train an AI system from a complete and objective set of examples.

Among the steps needed to make AI more fair is eliminating bias in datasets and establishing systems to guarantee that all demographics are treated fairly and have equal access to results. Several factors contribute to a sense of fairness, including equality of opportunity, predictive parity, and demographic parity. To avoid unfair treatment of applicants because of their gender, race, or ethnicity, for instance, AI-driven hiring tools must adhere to the principle of nondiscrimination. Using several fairness criteria might lead to inconsistencies or conflicts with other AI performance goals, making the pursuit of fairness fraught with complexity and trade-offs. The sorts of dataset bias that can impair AI models are covered in Table 13.2.

Efforts to address dataset bias and fairness should not be seen as a one-and-done operation; they should be monitored and adjusted continuously throughout the AI model's lifetime. Evaluating the accuracy of AI forecasts in practical settings is an important part of this process, as is performing comprehensive analyses of the training data to detect biases. For AI applications to be fair and dataset bias properly addressed, interdisciplinary teams of data scientists, ethicists, domain experts, and affected communities must work together.

TABLE 13.2
Types of dataset bias

Type	Impact on Model Performance	Detection Methods
Sampling Bias	Leads to inaccurate or skewed results	Statistical analysis, demographic checks
Measurement Bias	Produces unreliable or invalid models	Consistency checks, validation studies
Labeling Bias	Causes model to learn incorrect patterns	Cross-validation, annotator agreement
Observer Bias	Introduces subjective errors	Blind data collection, training sessions
Selection Bias	Excludes important data, reducing generalizability	Random sampling, review of selection criteria
Confirmation Bias	Skews results toward expected outcomes	Hypothesis-blind analysis
Recall Bias	Results in inaccurate historical data	Cross-referencing, verification methods
Reporting Bias	Hides important data, leading to biased conclusions	Comprehensive data checks
Nonresponse Bias	Limits dataset completeness	Follow-up surveys, response rate analysis
Exclusion Bias	Removes critical data, affecting results	Review of exclusion criteria
Survivorship Bias	Overestimates success rates	Historical data analysis, inclusion of all data points
Funding Bias	Results are swayed by funding sources	Disclosure of funding sources
Publication Bias	Skews scientific understanding	Literature reviews, pre-registration of studies

13.3.2 ETHICAL DATA COLLECTION PRACTICES

Data, frequently called the "lifeblood" of AI, tremendously influences how AI models act and the results they produce. Concerns about data bias and discrimination and questions of privacy and consent are just a few ethical challenges that could arise throughout the data collection process. Fairness, openness, and respect for individual rights are the guiding principles of ethical data-collecting procedures, which seek to overcome these challenges.

The idea of informed consent is crucial to ethically collecting data. According to this principle, people should be able to make an informed decision about whether or not to consent after being fully informed about the intended use of their data. This implies that people should easily find and understand information regarding data collection's goals, procedures, and risks. Also, people should be able to revoke their permission and be given freely, without manipulation or pressure. Building trust between data collectors and individuals and meeting regulatory criteria like the General Data Protection Regulation (GDPR) are achieved through strong consent systems.

Additionally, protecting privacy and confidentiality is integral to ethical data-collecting techniques. This necessitates taking precautions to prevent sensitive data loss, misuse, or alteration. Data anonymization, encryption, and access controls are ways that information can be protected from prying eyes and other unwanted outcomes like prejudice and identity theft. A company's data management policies and procedures should spell out specifics like how long data should be kept, how it should be shared, and how it should be disposed of.

13.3.3 ENSURING REPRESENTATIVENESS IN TRAINING DATA

Unfair results and damaging decisions may come from prejudiced and skewed models due to a lack of representation. Several methods and approaches are used to overcome this obstacle during the data gathering, preprocessing, and training of the model's stages.

Rigid sampling strategies that attempt to capture the diversity of the target population can be one way to ensure that training data is representative. This necessitates gathering information from several sources and checking that the dataset sufficiently represents all pertinent demographics, traits, and situations. Also, by adding variations and perturbations to current samples, data augmentation techniques can artificially improve the training data's diversity. In order to make more accurate predictions and avoid unjust results, it is important to increase the diversity of the dataset so that AI models can generalize to new data.

It is crucial to continuously monitor and evaluate AI models in production settings to detect and resolve any inconsistencies or biases that may develop over time. As part of this process, it is necessary to regularly assess how well the model performs across various demographic groups and how different stakeholders will be affected by its predictions. Organizations can work toward AI systems that are fair and inclusive by continuously improving and upgrading training data and model algorithms using real-world feedback.

13.4   AI GOVERNANCE AND COMPLIANCE

AI regulations are complicated and often changing. As AI develops rapidly, govern-ments and international organizations need help establishing efficient governance mechanisms. The quick evolution of AI technology, the necessity for international cooperation, and the borderless character of AI all work against the creation of thor-ough and efficient regulatory frameworks [17].

Important steps include preventing discrimination and bias, considering how AI will affect society and the environment, and implementing safeguards to make AI systems accountable and auditable [17]. Nevertheless, there is still a long way to go before we can turn theoretical ethical concepts into workable procedures for AI gov-ernance [18]. Given the varied effects of AI across various sectors, it is also impor-tant to establish industry-specific rules [17].

Before implementing AI into their operations, firms must establish internal gov-ernance systems. As part of this effort, multi-tiered AI governance systems are being considered, with each level covering a different aspect of AI development, regulation, oversight, and ethical and legal considerations. Including AI governance in preexisting organizational governance frameworks, including data governance, corporate governance, and IT governance, is critical. Importantly, "intersubjec-tively recognized rules that define, constrain, and shape expectations about the fun-damental properties of an artificial agent" (AR) should be operationalized as AI governance [18].

13.5   ETHICAL CHALLENGES IN AI APPLICATIONS

13.5.1   Autonomous Systems and Moral Agency

In particular, autonomous systems equipped with AI bring up serious concerns regarding free will. Historically, the ability of people to behave morally has been associated with moral agency. This ability is most commonly seen when people make conscious decisions based on their code of ethics and moral reasoning. The idea of moral agency, however, broadens to encompass robots and humans as AI systems gain more and more autonomy and decision-making power.

The subject of whether or not autonomous systems may be held ethically account-able for their deeds is central to the discussion. The basis of AI systems' operations are algorithms, data inputs, and specified objectives, as opposed to the consciousness and subjective experiences that human agents possess. They cannot experience moral feelings, have any intentional behavior, or be conscious in the conventional sense. So, there are a lot of philosophical and practical problems with giving autonomous systems moral agency.

Philosophers, engineers, and ethicists are debating many important questions about moral agency and autonomous systems. The issue of responsibility comes up first. When AI systems make harmful or unethical actions, who is to blame? Is the AI system to blame or the people who created, programmed, or ran it? When AI works in constantly changing and unpredictable circumstances, with little human supervi-sion, assigning blame becomes very difficult.

Second, the question of explainability and transparency comes up. The only way for autonomous systems to prove they are morally responsible is if they are open about how they make decisions. In order to help people assess the ethical consequences of AI decisions, explainable AI (XAI) methods try to shed light on how these systems make their judgments. However, it is still not easy to get deep learning models—which make decisions via opaque decision-making processes—to be very transparent.

Whether AI goals are compatible with ethical values is a source of concern. The goals or utility functions upon which autonomous systems are designed to work are not necessarily congruent with human ideals or accepted social standards. Unanticipated repercussions or ethical quandaries could arise if AI systems are not designed and monitored carefully to avoid prioritizing optimization or efficiency over ethical concerns.

13.5.2 JOB DISPLACEMENT AND SOCIOECONOMIC IMPACTS

AI's automation of formerly human-only jobs can cause changes in employment patterns or even eliminate jobs [19]. Huge unemployment and economic instability could result from sudden changes in the job market caused by AI's mass displacement of human workers (Figure 13.2). No industry or geographic area is immune to the possibility of employment loss due to AI. The International Monetary Fund found that in developed countries, new job opportunities, especially in AI development and associated industries, can be created by AI, making it impossible to forecast the overall effect. According to the World Economic Forum, worldwide companies plan to add 69 million jobs by 2027. However, they also intend to eliminate 83 million jobs, meaning there will be a net loss of 14 million roles [20]. Beyond individual employment, the socioeconomic ramifications of AI-driven job displacement are far-reaching. The jobs replaced by AI typically involve physical or routine duties, whereas those created by AI typically demand specialized knowledge and abilities, which can worsen income inequality [20].

Social safety nets can become overwhelmed, and people can become unemployed for longer due to this skills gap. Investments in education, retraining programs, and supportive measures for displaced workers are important proactive policy initiatives to counteract these harmful consequences. Effective cooperation among governments, schools, companies, and individuals is paramount when tackling the problems and seizing the opportunities that come with AI in the workplace.

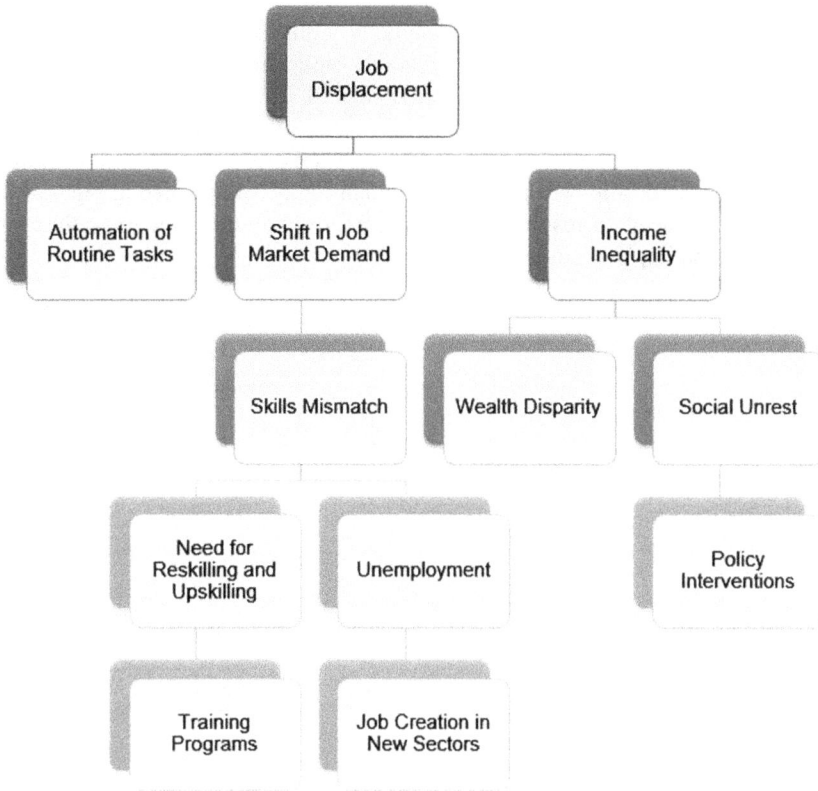

FIGURE 13.2   Effects of job displacement on economic and social processes

13.5.3   AI AND HUMAN RIGHTS

One side of the coin claims that AI can help advance human rights by expanding opportunities in healthcare, education, and information access [21, 22]. For instance, medical diagnostics driven by AI have the potential to enhance healthcare outcomes, and search engines powered by AI can increase people's access to information [21]. However, there are serious concerns and dangers to human rights that AI presents as well [21–23].

Decisions made by AI systems might be biased against specific groups if the data used to train them reflects societal biases. The right to nondiscrimination and equality could be violated due to this. The effect of AI on personal data and privacy is another important concern. There are worries about the possibility of monitoring, manipulation, and losing control over one's personal information when AI systems gather and handle large volumes of personal data [21–23].

AI also brings up problems with due process and access to appropriate remedies. There are worries over the lack of accountability and transparency when AI systems are utilized in decision-making situations with high stakes, including criminal sentences or loan approvals. This could make it hard for people to fight back against decisions that hurt them. The employment of AI in policing and combat also raises questions regarding individual safety and the right to life. The intricate and multi-faceted relationship between AI and human rights is illustrated in Figure 13.3, highlighting the need to resolve these concerns to guarantee fair and ethical AI development and implementation.

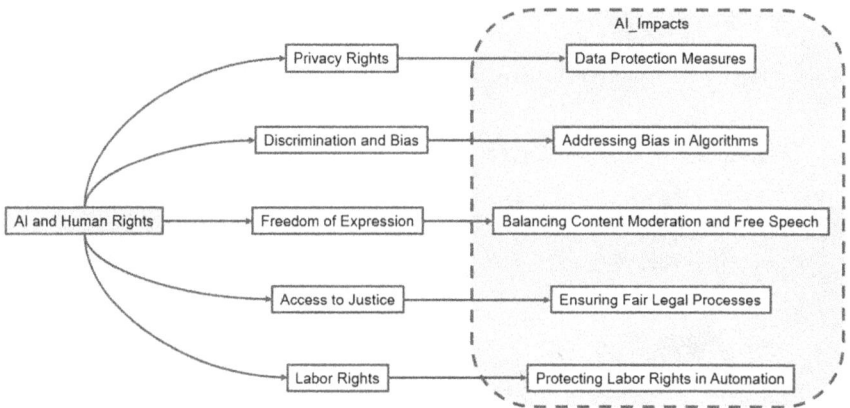

FIGURE 13.3   Ethical dimensions of AI and human rights

13.6   TOOLS AND TECHNOLOGIES FOR ETHICAL AI DEVELOPMENT

13.6.1   EXPLAINABLE AI (XAI) TECHNIQUES

If the underlying AI models or data change, the explanations generated by XAI methods using the training data could be more accurate and updated. The effectiveness and utility of XAI systems depend on their ability to automatically update explanations in response to changes in the underlying models or data. Maintaining the efficiency of XAI approaches and delivering credible explanations in dynamic situations requires adapting them to handle altering data distributions. When it comes to large-scale AI systems, scalability is key. XAI techniques that are effective on controlled or small-scale datasets might need help to handle the complexity of the models and the sheer volume of data involved [24].

The requirement for interpretability and transparency in AI decision-making processes drives the development of XAI techniques [25]. The need for explanations that help establish confidence and comprehension in AI systems is rising with their complexity and prevalence. Each XAI approach, from model-agnostic to model-specific, has advantages and disadvantages [26]. The application domain, AI model type, and desired explainability level dictate the XAI technique.

Continuous explanation updates, adaptability to shifting data distributions, and scalability are some of the limitations and obstacles that current XAI systems need to overcome. These issues will shape the future of XAI research [24–27]. New human-AI interfaces are also required, so AI systems can perceive context and apply conceptual knowledge. It is anticipated that the global XAI community will play a pivotal role in this domain, propelling the creation of effective XAI models that meet specific requirements for causal efficiency and user pleasure in a particular application setting [26].

13.6.2 Fairness Assessment Tools

These instruments aim to test how well AI algorithms account for varying degrees of inequality in terms of age, gender, ethnicity, and socioeconomic position. In order to find instances of prejudice or bias, fairness evaluation tools examine the results of AI models. This allows developers to fix these problems before the models are deployed. These tools enable organizations to make informed decisions about the use and impact of AI systems by providing quantitative insights into their fairness and equity using various metrics and techniques.

As an approach, assessment methods for fairness sometimes use fairness measures, which quantify the difference in outcomes experienced by various demographic groups. Measures like demographic parity and differential impact, which quantify the difference in outcomes between groups, are statistical tools that fall under this category of metrics. Fairness evaluation tools can help developers alter algorithms to reduce prejudice by applying these metrics to AI models and finding potential regions of bias. Furthermore, fairness assessment tools frequently use techniques like sensitivity analysis and counterfactual testing to investigate how algorithmic decisions may affect different demographic groups further.

These tools help firms show they are committed to ethical AI practices and regulatory compliance by giving developers quantifiable estimates of algorithmic fairness. Fostering responsible AI innovation and establishing confidence among users and stakeholders requires AI systems to be transparent. Figure 13.4 summarizes the tools used to evaluate AI fairness.

13.6.3 Privacy-Preserving AI Technologies

These technologies are meant to enable the design and implementation of AI systems while guaranteeing private user data protection. One tactic is to use differential privacy, which adds noise to the data to hide individual information while enabling effective analysis at the aggregate level. Companies that respect privacy this way can use large datasets to train AI models without compromising the protection of private personal data.

Among the many methods for protecting privacy in AI is homomorphic encryption. By using this approach, computations on encrypted data can be done without first decrypting it. This guarantees that the data confidentiality is preserved through the analysis process. This allows several companies to work together on AI projects and securely exchange data without disclosing important information to people who are not supposed to see it. Federated learning has also become a workable answer to privacy-preserving AI, especially where legislative or privacy issues prevent data from being centralized. This covers circumstances when centralization of the data is impossible. Federated learning uses locally executed model training on scattered servers or devices. Raw data will remain decentralized and private as only updated parameters are pooled.

It is impossible to overstate the significance of privacy-protecting technology, particularly because AI applications are still in demand in various sectors. These technologies support confidence and openness in AI systems and safeguard personal sensitive data.

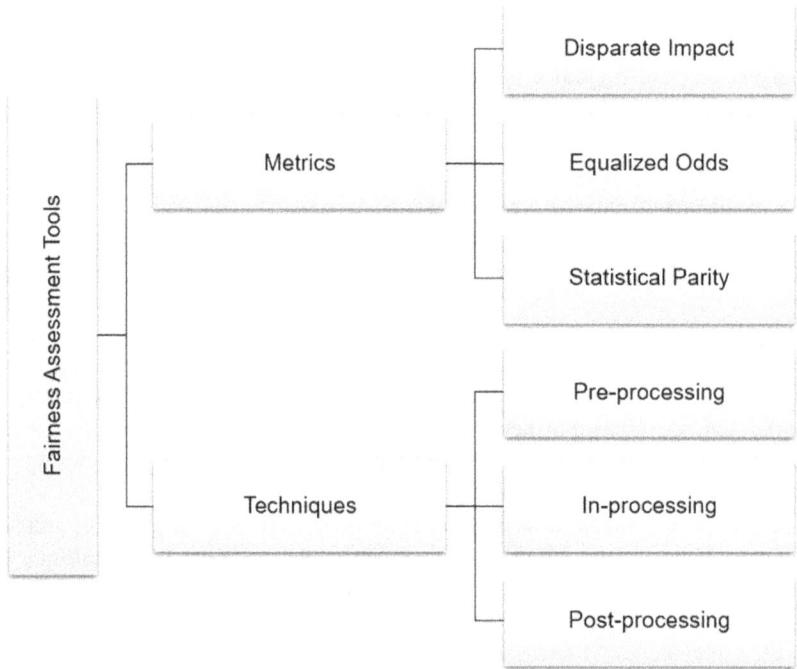

FIGURE 13.4   Overview of fairness assessment tools

13.6.4   Managing Ethical Challenges in Data-Driven AI Systems

Many machine learning algorithms are opaque and unpredictable, which presents one of the biggest problems. This is such that no users, deployers, or developers can completely understand or predict how the system will react to the given inputs [28]. This lack of interpretability and accountability raises questions about prejudice, justice, and potential unanticipated negative consequences [29, 30]. Individual privacy rights must be guaranteed by AI systems handling sensitive personal data, such as those used in healthcare applications, adhering to strict data protection regulations. AI has to be built and applied ethically in delicate domains to build public trust and avoid harm to people and communities [30]. Table 13.3 lists important ethical concerns in data-driven AI systems.

The increasingly extensive effects of AI on society raise ethical issues, such as how technology affects democracy, justice, and employment [28]. Concerns exist over how AI can exacerbate inequalities or lead to new discrimination [29]. This is so because AI is spreading more and more everywhere. Ethical governance frameworks must be in place to ensure that AI is created and applied in a way that benefits society at large rather than benefiting some groups at the expense of others.

TABLE 13.3
Overcoming ethical obstacles of data-driven AI systems

Ethical Challenge	Risks	Solutions	Technologies/Practices
Data Privacy	Data breaches, misuse	Encryption, access controls	GDPR compliance, anonymization
Bias and Fairness	Discrimination, biased outcomes	Diverse datasets, bias detection	Fairness-aware algorithms, regular audits
Transparency	Opaque algorithms	Explainable AI (XAI), clear documentation	Model interpretability frameworks
Accountability	Diffuse responsibility	Clear accountability structures	Documentation, regular reviews
Consent	Lack of user awareness	Simplified consent forms	User-friendly opt-in/opt-out mechanisms
Security	Adversarial attacks	Robust security measures	Penetration testing, adversarial training
Data Quality	Inaccurate, incomplete data	Data cleaning, validation	Continuous data quality monitoring
Human Oversight	Over-reliance on AI	Hybrid decision systems, regular reviews	Clear escalation paths
Ethical Data Collection	Intrusive methods, privacy issues	Transparent practices, respect for privacy	Adherence to ethical guidelines
Impact Assessment	Unintended consequences	Regular impact assessments	Stakeholder consultations
Regulatory Compliance	Non-compliance risks	Compliance audits, legal consultations	Proactive regulatory engagement
Ethical Use of AI Outputs	Misuse of outputs	Clear usage policies, ethical guidelines	Regular ethical training for users
Environmental Impact	High energy consumption	Energy-efficient algorithms	Renewable energy sources, recycling practices

13.7 AI SECURITY

13.7.1 ADVERSARIAL ATTACKS ON AI SYSTEMS

AI systems are in danger of adversarial attacks, compromising their security and dependability. This is particularly true in applications as crucial as driverless cars, healthcare diagnostics, and financial fraud detection [31–34]. These attacks involve a technique of manipulating input data such that, while remaining invisible to humans, AI models misclassify or produce incorrect outputs [31, 33, 34]. Significant safety and ethical issues arise when AI systems become more prevalent in real-world contexts [32, 33].

Aggressive attacks have been launched using several techniques. Among these techniques are incorporating subtle perturbations into the input data, developing harmful inputs intended to fool AI models, and using algorithmic weaknesses [34, 35]. Untargeted attacks aim to cause the model to misclassify the input into any incorrect category. In contrast, targeted attacks aim to force a model to categorize an input into a specific category. Adversarial assaults can target image classifiers, natural language processing models, and reinforcement learning agents, among other AI system components [34].

Researchers have improved the fundamental characteristics of adversarial attacks and their transferability across models [32, 34]. Single-step attack techniques are often more portable than multi-step assault techniques, which qualifies them for black-box attacks [32]. Conversely, robustness against some attack techniques may be increased by adversarial training, which is the deliberate training of a model on hostile scenarios [32]. Currently, research in this area aims to provide algorithms that can assess computation and explain the results of AI models, thereby providing verifiable protections against adversarial attacks [33, 35].

13.7.2 SECURING AI MODELS AND DATA

Implementing differential privacy is an important part of protecting AI models. This technique adds noise to the input so that individual information cannot be extracted, but the model's overall pattern recognition capabilities are preserved [36]. This method is critical for preserving sensitive data in various contexts, particularly in the healthcare industry, where patient privacy is paramount. By training models across dispersed devices without exchanging raw data, federated learning has emerged as a viable way to enhance data security, thus preserving privacy and secrecy [37].

Strengthening AI systems' capacity to protect users' privacy requires incorporating encryption and safe multiparty computing [2]. Ensuring the security and confidentiality of data throughout the AI lifecycle is a top priority for professionals adept in these areas. Compliance with quality standards, rules, and regulations allows firms to establish a strong data governance system that protects against data breaches and unauthorized access. This foresightful method promotes openness and responsibility in handling sensitive data while simultaneously strengthening data security.

Efficient processing and analysis approaches are necessary to ensure data integrity and security in real-time issues where the velocity and volume of data provide substantial difficulties [38]. Tackling these difficulties calls for a comprehensive strategy that integrates state-of-the-art encryption techniques, safe procedures for transferring data, and constant threat detection and mitigation through monitoring.

13.8 FUTURE DIRECTIONS IN AI ETHICS AND IT

The issue of ensuring that AI systems uphold human rights and human values is becoming increasingly important in the fast-growing field of AI ethics. Researchers are investigating many approaches to the creation of AI, including ethical issues. These techniques include using inclusive and diverse data sets, open decision-making procedures, and accountability systems. The need for ethical AI and programming is brought to light by the possible dangers connected to biased or unaccountable AI systems, which may exacerbate societal inequalities or compromise human privacy and freedom.

One instance of how, as AI technologies get more sophisticated, ethical issues surrounding them will become increasingly crucial is the development of Artificial General Intelligence (AGI) [39]. AGI raises concerns about the viability of autonomous decision-making and the need for safeguards to ensure that such systems adhere to moral principles and human values [39]. The ethical ramifications of advanced AI, which include the potential for job displacement, the need for openness and explainability in AI decision-making, and the need to create AI systems that respect human dignity and autonomy, are a challenge for researchers to grasp [39, 40].

To develop morally sound AI that benefits society, many stakeholders—including politicians, business executives, and civil society members—must cooperate [40, 41]. One set of rules for the moral use of AI in the healthcare sector was developed by the World Health Organization (WHO). These values stress the need for responsibility, transparency, and human rights respect [41]. In a similar spirit, the European Commission has underlined the need for the ethical advancement of AI, pointing out that although AI has the potential to democratize access to vital industries, it also presents issues that require coordinated international action [41]. The outcomes of these projects show how a coordinated strategy for creating ethical AI is needed, one that finds a balance between the benefits of AI and the need to protect human rights and dignity [40, 41].

In the future, research in the field of AI ethics should focus on the creation of useful guidelines and principles for the creation of ethical AI, as well as the analysis of the ethical ramifications of new developments in the field [39–42]. What is mentioned in this sentence are, for instance, "this includes examining the impact of AI on meaningful work," "ensuring that AI systems are designed to promote human flourishing and dignity," and "developing mechanisms for accountability and transparency in AI decision making." When ethical issues are given high priority in the creation of AI, we may ensure that emerging technologies will benefit society at large while upholding each individual's rights and dignity.

REFERENCES

1. Ali, M.A., Yap, N.K., Ghani, A.A.A., Zulzalil, H., Admodisastro, N.I. and Najafabadi, A.A., A systematic mapping of quality models for AI systems, software and components. *Applied Sciences*, 2022. 12(17): p. 8700.
2. Sarker, I.H., AI-based modeling: Techniques, applications and research issues towards automation, intelligent and smart systems. *SN Computer Science,* 2022. 3(2): p. 158.
3. Cabrero-Daniel, B., *AI for agile development: A meta-analysis.* arXiv preprint arXiv:2305.08093, 2023.
4. Dingsøyr, T., Nerur, S., Balijepally, V. and Moe, N.B., *A decade of agile methodologies: Towards explaining agile software development.* 2012: Elsevier. pp. 1213–1221.
5. Rasnacis, A. and S. Berzisa, Method for adaptation and implementation of agile project management methodology. *Procedia Computer Science,* 2017. 104: pp. 43–50.

6. Žlahtič, B., Završnik, J., Blažun Vošner, H., Kokol, P., Šuran, D. and Završnik, T., Agile machine learning model development using data canyons in medicine: A step towards explainable artificial intelligence and flexible expert-based model improvement. *Applied Sciences*, 2023. 13(14): p. 8329.

7. Karamitsos, I., S. Albarhami, and C. Apostolopoulos, Applying DevOps practices of continuous automation for machine learning. *Information*, 2020. 11(7): p. 363.

8. De Silva, D. and D. Alahakoon, An artificial intelligence life cycle: From conception to production. *Patterns*, 2022. 3(6): pp. 1–13.

9. Steidl, M., M. Felderer, and R. Ramler, The pipeline for the continuous development of artificial intelligence models—Current state of research and practice. *Journal of Systems and Software*, 2023. 199: p. 111615.

10. Azad, N. and S. Hyrynsalmi, DevOps critical success factors—A systematic literature review. *Information and Software Technology*, 2023. 157: p. 107150.

11. Sriraman, G., A machine learning approach to predict DevOps readiness and adaptation in a heterogeneous IT environment. *Frontiers in Computer Science*, 2023. 5: p. 1214722.

12. Yu, L., Alégroth, E., Chatzipetrou, P. and Gorschek, T.,Automated NFR testing in continuous integration environments: A multi-case study of Nordic companies. *Empirical Software Engineering*, 2023. 28(6): p. 144.

13. Fernández González, D., Rodríguez Lera, F.J., Esteban, G. and Fernández Llamas, C., Secdocker: Hardening the continuous integration workflow: Wrapping the container layer. *SN Computer Science,* 2022. 3: pp. 1–13.

14. Bertolini, R., S.J. Finch, and R.H. Nehm, Enhancing data pipelines for forecasting student performance: Integrating feature selection with cross-validation. *International Journal of Educational Technology in Higher Education*, 2021. 18: pp. 1–23.

15. Pressman, S.M., Borna, S., Gomez-Cabello, C.A., Haider, S.A., Haider, C. and Forte, A.J, AI and ethics: A systematic review of the ethical considerations of large language model use in surgery research, in *Healthcare*. 2024, MDPI. p. 825.

16. Wei, G. and H. Niemi, Ethical guidelines for artificial intelligence-based learning: A transnational study between China and Finland. *Learning: Designing the Future*, 2023: p. 265.

17. Walter, Y., Managing the race to the moon: Global policy and governance in Artificial Intelligence regulation—A contemporary overview and an analysis of socioeconomic consequences. *Discover Artificial Intelligence*, 2024. 4(1): p. 14.

18. Mäntymäki, M., Minkkinen, M., Birkstedt, T. and Viljanen, M., Defining organizational AI governance. *AI and Ethics*, 2022. 2(4): pp. 603–609.

19. Idrisi, M.J., D. Geteye, and P. Shanmugasundaram, Modeling the complex interplay: Dynamics of job displacement and evolution of artificial intelligence in a socio-economic landscape. *International Journal of Networked and Distributed Computing*, 2024. 12.: pp. 1–10.

20. Nnamdi, N., B.Z. Ogunlade, and B. Abegunde, An evaluation of the impact of artificial intelligence on socio-economic human rights: A discourse on automation and job loss. *Scholars International Journal of Law, Crime and Justice*, 2023. 6(10): pp. 508–521.

21. Kriebitz, A. and C. Lütge, Artificial intelligence and human rights: A business ethical assessment. *Business and Human Rights Journal*, 2020. 5(1): pp. 84–104.

22. Ashraf, C., Exploring the impacts of artificial intelligence on freedom of religion or belief online. *The International Journal of Human Rights*, 2022. 26(5): pp. 757–791.

23. Jarrahi, M.H., C. Lutz, and G. Newlands, Artificial intelligence, human intelligence and hybrid intelligence based on mutual augmentation. *Big Data & Society*, 2022. 9(2): p. 20539517221142824.

24. Yang, W., Wei, Y., Wei, H., Chen, Y., Huang, G., Li, X., Li, R., Yao, N., Wang, X., Gu, X. and Amin, M.B., Survey on explainable AI: From approaches, limitations and applications aspects. *Human-Centric Intelligent Systems*, 2023. 3(3): pp. 161–188.

25. Dwivedi, R., Dave, D., Naik, H., Singhal, S., Omer, R., Patel, P., Qian, B., Wen, Z., Shah, T., Morgan, G. and Ranjan, R., Explainable AI (XAI): Core ideas, techniques, and solutions. *ACM Computing Surveys*, 2023. 55(9): pp. 1–33.

26. Holzinger, A., Saranti, A., Molnar, C., Biecek, P. and Samek, W., Explainable AI methods- a brief overview, in *International workshop on extending explainable AI beyond deep models and classifiers*. 2022, Springer. pp. 13–38

27. Hulsen, T., Explainable Artificial Intelligence (XAI): Concepts and challenges in healthcare. *AI*, 2023. 4(3): pp. 652–666.

28. Stahl, B.C., Ethical issues of AI, in *Artificial Intelligence for a better future: An ecosystem perspective on the ethics of AI and emerging digital technologies*, Springer Nature. 2021: pp. 35–53.

29. Heyder, T., N. Passlack, and O. Posegga, Ethical management of human-AI interaction: Theory development review. *The Journal of Strategic Information Systems*, 2023. 32(3): p. 101772.

30. Mohammad Amini, M., Jesus, M., Fanaei Sheikholeslami, D., Alves, P., Hassanzadeh Benam, A. and Hariri, F., Artificial intelligence ethics and challenges in healthcare applications: A comprehensive review in the context of the European GDPR mandate. *Machine Learning and Knowledge Extraction*, 2023. 5(3): pp. 1023–1035.

31. Xu, H., Ma, Y., Liu, H.C., Deb, D., Liu, H., Tang, J.L. and Jain, A.K., Adversarial attacks and defenses in images, graphs and text: A review. *International Journal of Automation and Computing*, 2020. 17: pp. 151–178.

32. Kurakin, A., I. Goodfellow, and S. Bengio, *Adversarial machine learning at scale.* arXiv preprint arXiv:1611.01236, 2016.

33. Zbrzezny, A.M. and A.E. Grzybowski, Deceptive tricks in artificial intelligence: Adversarial attacks in ophthalmology. *Journal of Clinical Medicine*, 2023. 12(9): p. 3266.

34. Ren, K., Zheng, T., Qin, Z. and Liu, X., Adversarial attacks and defenses in deep learning. *Engineering*, 2020. 6(3): pp. 346–360.

35. Chang, C.L., Hung, J.L., Tien, C.W., Tien, C.W. and Kuo, S.Y., Evaluating robustness of AI models against adversarial attacks, in *Proceedings of the 1st ACM Workshop on Security and Privacy on Artificial Intelligence*. 2020.

36. Aldoseri, A., K.N. Al-Khalifa, and A.M. Hamouda, Re-thinking data strategy and integration for artificial intelligence: Concepts, opportunities, and challenges. *Applied Sciences*, 2023. 13(12): p. 7082.

37. Khalid, N., Qayyum, A., Bilal, M., Al-Fuqaha, A. and Qadir, J., Privacy-preserving artificial intelligence in healthcare: Techniques and applications. *Computers in Biology and Medicine*, 2023: p. 106848.

38. Rahmani, A.M., et al., Artificial intelligence approaches and mechanisms for big data analytics: A systematic study. *PeerJ Computer Science*, 2021. 7: p. e488.

39. Kazim, E. and A.S. Koshiyama, A high-level overview of AI ethics. *Patterns*, 2021. 2(9): pp. 1–12.

40. Bankins, S. and P. Formosa, The ethical implications of artificial intelligence (AI) for meaningful work. *Journal of Business Ethics*, 2023. 185(4): pp. 725–740.

41. Bouderhem, R., Shaping the future of AI in healthcare through ethics and governance. *Humanities and Social Sciences Communications*, 2024. 11(1): pp. 1–12.

42. Roche, C., P. Wall, and D. Lewis, Ethics and diversity in artificial intelligence policies, strategies and initiatives. *AI and Ethics*, 2023. 3(4): pp. 1095–1115.

Index

advanced persistent threats 38
algorithmic bias 67, 83–85, 113–114, 121

bias detection 13, 112, 114–115

cognitive biases 81–82
conflicts of interest 21–22, 105, 155, 159, 164, 172, 174–175
cyberattacks 38–39, 45, 48, 68, 73, 78, 121, 139, 190

data privacy 2, 5, 10, 14–15, 28, 62–63, 109, 139, 147, 161, 170, 189–190, 211
deontology 2–3, 16, 18, 22, 33, 143, 153
digital citizenship 4, 27, 52, 54, 148–149
digital forensics 55, 57–58
digital freedom 77–78

ethical awareness 5, 14, 19, 40–41, 45, 181–182
ethical branding 102
ethical codes 14, 58, 106
ethical gray areas 9
ethical hacking 41–43, 195
ethical pluralism 4
ethical principles 4, 6–8, 15, 24, 26, 97, 105–109, 121, 129, 157, 166, 176, 202
ethical reporting 31–32
ethical theories 2, 6, 16

individual privacy 28, 54, 66, 68–69, 121, 170, 176, 211
intellectual property 12, 25, 27, 34, 40, 90, 191

Markkula Center Framework 8

national security 40, 56, 67–68, 176, 194
Nine-Box Model 7

offshoring 23–27
online behavior 40, 54, 88, 148–149
open communication 10, 14–15, 20, 24, 26, 33, 49, 54, 56, 96, 110, 122, 144
open source 96–98
outsourcing 23–27

Potter Box Model 7
privacy rights 57, 62, 67–69, 74, 159, 170, 185, 190, 194, 211
Protection and Electronic Documents Act 188

SAD Formula 7
security measures 1, 4, 10, 41, 44, 46, 49, 69, 71–72, 110, 117, 121, 139, 141, 159–160, 169, 194, 211
social contract theory 4, 18
social impact 39, 128
sustainable technology 129

tech industry 191

unauthorized access 28, 40–41, 43, 121, 140, 147, 169, 195, 212
user needs 85, 147
utilitarianism 2, 16, 18, 22, 153, 155

virtue ethics 3, 16, 18, 22, 33, 143, 153, 155

Wall Street Journal Model 8

For Product Safety Concerns and Information please contact our EU
representative GPSR@taylorandfrancis.com
Taylor & Francis Verlag GmbH, Kaufingerstraße 24, 80331 München, Germany